Research Methods
in
Plant Science

Research Methods in Plant Science

Vol. 1

Dr. Chandra P. Singh

AGROTECH PRESS
JAIPUR ▪ NEW DELHI

AGROTECH PRESS

Regd. Off. "Bony Residency", Gate No. 2
Opp. Tilak Public School, Vishweswariya Nagar,
Gopalpura Road, Jaipur - 302018, India
Ph: 0141 - 2761280, Telefax: 2761381
E-mail : agrotechpress@gmail.com

Delhi Office: 102, 1st Floor, Satyam House,
4327/3, Ansari Road, Darya Ganj,
New Delhi - 110002, India
Ph: 011-45652440
visit us: www.agrotechpress.com

Distributed By
KEYA INTERNATIONAL PRESS LLP
E-mail: keyainternationalpressllp@gmail.com

Published in India, 2025

ISBN : 9789384568542 (2 Vols. Set)

Typeset by Thomson Press

Printed and bound in India by Thomson Press Ltd.,
New Delhi

Contents

Preface

Allelopathy refers to a mechanism by which a plant interacts with another or a neighbouring organism by chemical compounds, which often have inhibitory effects on the seed germination or growth of the other plant. Allelochemicals are non-nutritional secondary metabolites produced by plants that have stimulatory or inhibitory effects upon the growth, health, behaviour, or population biology of neighbouring organisms. In plant–plant allelopathy, the more common inhibitor effects are visible upon such plant functions as respiration; photosynthesis; water balance and stomatal function; stem conductance of water; xylem element flux; membrane permeability; cell division and development; protein synthesis; and enzyme activity alteration. Within plants, allelochemicals may be distributed broadly among organs such as seeds, flowers, pollen, leaves, stems, and roots, or sometimes found in just one or two such locations. Allelochemicals can get into the environment in several ways; these include volatisation, exudation from roots, leaching from plants by rain and decomposition of residues. The production of allelochemicals often decreases as the plant ages. The effects of this can be stimulatory, but often they are inhibitory.

Hence allelopathy is an important factor in plant ecology, because it can affect the distribution patterns of plants and at the same time often decreases the biodiversity of the ecosystem. Allelopathy has also significant applications for example in the fields of agriculture and forestry, because allelochemicals can for example be used to protect crop plants from weeds. Allelopathy is different from, even though their effects are often non-separable. For example a small allelopathic disadvantage can place a plant in a worse position if competition is also taking place, because its growth can be prevented by the other plant, which produces the chemicals. However the production of allelochemicals exhausts also the producing plant, and thus plants often produce significant amounts of allelochemicals only when they are under stress. The discipline of allelopathy has been experiencing extraordinary growth in research and there is a great deal of interest in seeking alternative environmentally friendly methods of weed control in agriculture, ways to deal with replant problems and soil sickness in horticulture, and competition control in young plantations. Research and application of allelopathy can help find some of the desirable alternatives. The deepening of our understanding of allelochemicals in plants holds much promise for widening our knowledge of plant ecosystems. Allelopathy is being increasingly recognized as a contributing explanation for such things as plant dominance, succession, formation of communities, climax vegetation, crop productivity, and exotic plant invasion.

Our understanding in allelopathy mechanisms has increased significantly with use of recently developed sophisticated tools and techniques in biochemistry, molecular biology and genetic engineering. Researchers in allelopathy are taking advantage of these rapidly expanding fields of molecular biology in understanding the complex plant–plant and plant-microbe interactions in seeking solutions to agricultural problems. Allelopathy holds out hope for improvements in crop production through such means as discovering eco-friendly herbicides with new sites of action, harmless to crops but toxic to weeds, and without formation of dangerous residues. Also, the genetic manipulation of crop germplasm to produce crops which carry inwardly all the desirable features of a successful competitor to their weeds, is a strongly motivating goal which is presently being pursued around the world. There are two major challenges to researchers on allelopathy - to minimize the negative impacts of allelopathy on crop growth and yield, and to exploit allelopathic mechanisms as additional pest control or crop growth regulation strategies. Plant products, microbial products, or their synthetic analogs could comprise the next generation of pesticides and growth regulaators. Joint efforts of chemists, plant physiologists, microbiologists, ecologists, and perhaps others, will be required to achieve maximum progress in this endeavour. Allelopathy research offers unlimited opportunities to solve practical agricultural problems and to contribute fundamental knowledge regarding the chemistry and biology of interspecific relationships.

This book is a timely effort to collect and synthesize the latest developments on allelopathy research with special emphasis on its application in agriculture and forestry. This book provides the reader relevant information about actual knowledge about the process of allelopathy, covering all aspects from the molecular to the ecological level. This will be highly useful guide for researchers and bio-scientists who will deal with one of the most strategic sciences to reduce the environmental pollution and to increase agricultural production in sustainable agriculture of the 21st century.

Dr. Chandra P. Singh

1

Introduction

Allelopathy is a biological phenomenon by which an organism produces one or more biochemicals that influence the growth, survival, and reproduction of other organisms. These biochemicals are known as allelochemicals and can have beneficial (positive allelopathy) or detrimental (negative allelopathy) effects on the target organisms. Allelochemicals are a subset of secondary metabolites, which are not required for metabolism (i.e. growth, development and reproduction) of the allelopathic organism. Allelochemicals with negative allelopathic effects are an important part of plant defence against herbivory.

Allelopathy is characteristic of certain plants, algae, bacteria, coral, and fungi. Allelopathic interactions are an important factor in determining species distribution and abundance within plant communities, and are also thought to be important in the success of many invasive plants.

The process by which a plant acquires more of the available resources (such as nutrients, water or light) from the environment without any chemical action on the surrounding plants is called resource competition. This process is not negative allelopathy, although both processes can act together to enhance the survival rate of the plant species.

History

The term allelopathy from the Greek-derived compounds *allelo-* and *-pathy* (meaning "mutual harm" or "suffering"), was first used in 1937 by the Austrian professor Hans Molisch in the book *Der Einfluss* *einer Pflanze auf die andere - Allelopathie* (The Effect of Plants on Each Other) published in German. He used the term to describe biochemical interactions that inhibit the growth of neighbouring plants, by another plant. In 1971, Whittaker and Feeny published a study in the journal *Science*, which defined allelochemicals as all chemical interactions among organisms. In 1984, Elroy Leon Rice in his monograph on allelopathy enlarged the definition to include all direct positive or negative effects of a plant on another plant or on micro-organisms by the liberation of biochemicals into the natural environment. Over the next ten years, the term was used by other researchers to describe broader chemical interactions between organisms, and by 1996 the International Allelopathy Society (IAS) defined allelopathy as "Any process involving secondary metabolites produced by plants, algae, bacteria and fungi that influences the growth and development of agriculture and biological systems." In more recent times, plant researchers have begun to switch back to the original definition of substances that are produced by one plant that inhibit another plant. Confusing the issue more, zoologists have borrowed the term to describe chemical interactions between invertebrates like corals and sponges.

Long before the term allelopathy was used, people observed the negative effects that one plant could have on another. Theophrastus, who lived around 300 BC noticed the inhibitory effects of pigweed on alfalfa. In China around the first century AD, the author of *Shennong Ben Cao Jing* described 267 plants that had pesticidal abilities, including those with allelopathic effects. In 1832, the Swiss botanist De Candolle suggested that crop plant exudates were responsible for an agriculture problem called soil sickness.

Allelopathy is not universally accepted among ecologists and many have argued that its effects cannot be distinguished from the competition which results when two (or more) organisms attempt to use the same limited resource, to the detriment of one or both. Allelopathy is a direct negative effect on one organism resulting from the input of substances into the environment by another. In the 1970s, great effort went into distinguishing competitive and allelopathic effects by some researchers, while in the 1990s others argued that the effects were often interdependent and could not readily be distinguished.

Examples of Allelopathy

The possible application of allelopathy in agriculture is the subject of much research. Current research is focused on the effects of weeds on crops, crops on weeds, and crops on crops. This research furthers the possibility of using allelochemicals as growth regulators and natural herbicides, to promote sustainable agriculture. A number of such allelochemicals are commercially available or in the process of large-scale manufacture. For example, Leptospermone is a purported thermochemical in lemon bottlebrush (*Callistemon citrinus*). Although it was found to be too weak as a commercial herbicide, a chemical analog of it, mesotrione (tradename Callisto), was found to be effective. It is sold to control broadleaf weeds in corn but also seems to be an effective control for crabgrass in lawns. Sheeja (1993) reported the allelopathic interaction of the weeds *Chromolaena odorata* (*Eupatorium odoratum*) and *Lantana camara* on selected major crops.

Many crop cultivars show strong allelopathic properties, of which rice (*Oryza sativa*) has been most studied. Rice allelopathy depends on variety and origin: Japonica rice is more allelopathic than Indica and Japonica-Indica hybrid. More recently, critical review on rice allelopathy and the possibility for weed management reported that allelopathic characteristics in rice are quantitatively inherited and several allelopathy-involved traits have been identified.

Many invasive plant species interfere with native plants through allelopathy. A famous case of purported allelopathy is in desert shrubs. One of the most widely known early examples was *Salvia leucophylla*, because it was on the cover of the journal *Science* in 1964. Bare zones around the shrubs were hypothesized to be caused by volatile terpenes emitted by the shrubs. However, like many allelopathy studies, it was based on artificial lab experiments and unwarranted extrapolations to natural ecosystems. In 1970, *Science* published a study where caging the shrubs to exclude rodents and birds allowed grass to grow in the bare zones. A detailed history of this story can be found in Halsey 2004.

Allelopathy has been shown to play a crucial role in forests, influencing the composition of the vegetation growth, and also provides an explanation for the patterns of forest regeneration. The black walnut *(Juglans nigra)* produces the allelochemical juglone, which affects some species greatly while others not at all. The leaf litter and root exudates of some *Eucalyptus* species are allelopathic for certain soil microbes and plant species. The tree of heaven, *Ailanthus altissima*, produces allelochemicals in its roots that inhibit the growth of many plants. The pace of evaluating allelochemicals released by higher plants in nature has greatly accelerated, with promising results in field screening.

Garlic mustard is an invasive plant species in North American temperate forests. Its success may be partly due to its excretion of an unidentified allelochemical that interferes with mutualisms between native tree roots and their mycorrhizal fungi.

A study of *Kochia scoparia* in northern Montana by two high school students showed that when *Kochia* precedes spring wheat (*Triticum aestivum*), it reduces the spring wheat's growth. Effects included delayed emergence, decreased rate of growth, decreased final height and decreased average vegetative dry weight of spring wheat plants. A larger study later showed that *Kochia* seems to exhibit allelopathy on various crops in northern Montana.

Corn gluten meal (CGM) is a natural preemergence weed control used in turfgrass, which reduces germination of many broadleaf and grass weeds.

Allelopathy in Plants

Plant allelopathy is all around us; yet, many people have never even heard of this interesting phenomenon. Allelopathy can have an adverse effect in the garden, resulting in reduced seed germination and plant growth. On the other hand, allelopathic plants may also be considered Mother Nature's own weed killer.

Allelpathy is a biological phenomenon where one plant inhibits the growth of another. How? Through the release of allelochemicals, certain plants can greatly affect the growth of other plants either in a good way or a bad one by leaching, decomposition, etc. In essence, plant allelopathy is used as a means of survival in nature, reducing competition from plants nearby.

Various parts of plants can have these allelopathic properties, from the foliage and flowers to the roots, bark, soil, and mulch. Most all allelopathic plants store their protective chemicals within their leaves, especially during fall. As leaves drop to the ground and decompose, these toxins can affect nearby plants. Some plants also release toxins through their roots, which are then absorbed by other plants and trees.

Common plants with allelopathic properties can be seen and include:

- English laurel (*Prunus laurocerasus*)
- Bearberry (*Arctostaphylos uva-ursi*)
- Sumac (*Rhus*)
- Rhododendron
- Elderberry (*Sambucus*)
- Forsythia
- Goldenrod (*Solidago*)
- Some types of fern

- Perennial rye
- Tall fescue
- Kentucky bluegrass

Trees are great examples of allelopathy in plants. For instance, many trees use allelopathy to protect their space by using their roots to pull more water from the soil so other plants cannot thrive. Some use their allelochemicals to inhibit germination or impede development of nearby plant life. Most allelopathic trees release these chemicals through their leaves, which are toxic once absorbed by other plants.

Black walnut is a prime example of this. In addition to its leaves, black walnut trees store allelopathic properties within their buds, nut hulls, and roots. The chemical responsible for its toxicity, called Juglone, remains in the soil around the tree and is most potent at the drip line, though the roots can spread out well beyond this. Plants most susceptible to the black walnut's toxicity include nightshade plants (tomatoes, peppers, eggplants, potatoes), azaleas, pines, and birch trees.

Other trees that are known to exhibit allelopathic tendencies include maple, pine, and eucalyptus.

Soil Allelochemicals: Sampling, Handling and Storage

Allelochemicals released into the soil by plants can affect the performance of interacting plant species. Allelochemicals can act as selective agents driving adaptation in associated plants species to cope with the compounds released by their "chemical neighbour". In this context, the soil microbial community can potentially affect the interaction between plants and plant chemicals as soil microorganisms may degrade allelochemicals after entering the soil. Most studies examining the effect of plant allelochemicals use bioassays where the effect of a chemical on seed germination and growth is tested in isolated petri dishes. By doing so these studies do not examine the effects of soil microorganisms.

Terpenes are the most common group of plant secondary compounds and are produced by a vast amount of plants – both herbs and trees. These compounds can have both inhibitory and stimulating effect on a number of associated organisms including plants, herbivores and microorganisms.

Many aromatic plants of the family *Lamiacae* produce monoterpenes as a main constituent of their essential oils. These monoterpenes are known for their antimicrobial activity mainly through their growth inhibitor effect on bacteria and fungi. However, it has been shown that some bacteria can decompose terpenes and use some of them as a carbon source. One genus well known for its production of monoterpenes is *Thymus*.

In some thyme species distinct chemotypes can be identified, where an individual plant often produces a single specific dominant monoterpene making up 60–80% of the essential oils that confer thyme plants their characteristic smell. Thyme monoterpenes enter the soil via leaf leachates, and are known to affect the performance of associated plants.

In general, the monoterpenes reduce seed germination and plant growth, but it has been shown that associated plants can adapt to their specific local thyme monoterpene. Studies on the impact of thyme monoterpenes on plant growth have been performed using non-sterile soil, either standard greenhouse soil where monoterpenes were manually added, or soil collected from natural sites.

The action of allelochemicals in soil needs their presence in the vicinity of the target plants. Ferulic, *p*-hydroxymandelic, *p*-hydroxybenzoic, and vanillic acids always had the lowest mobility ($Rf < 0.1$), whereas phenolic aldehyde and lactone (vanillin and coumarin) showed the highest mobility ($Rf > 0.5$). The *Rf* values of daidzein, 1α-angeloyloxycarotol, DIMBOA, and *m*-tyrosine ranged from 0.24 to 0.32. Binary mixtures of these allelochemicals led to an increase in mobility factors for selected combinations. Phospholipid fatty acid profiling indicated that there were different soil microbial communities in the segments containing allelochemicals residues in the developed TLC soil layer.

Colourimetric Reactions in Soil Allelochemicals Analysis

Colourimetric analysis is a method of determining the concentration of a chemical element or chemical compound in a solution with the aid of a colour reagent. It is applicable to both organic compounds and inorganic compounds and may be used with or without an enzymatic stage. The method is widely used in medical laboratories and for industrial purposes, e.g. the analysis of water samples in connection with industrial water treatment.

Equipment

The equipment required is a colourimetre, some cuvettes and a suitable colour reagent. The process may be automated, e.g. by the use of an AutoAnalyzer or by Flow injection analysis.

Non-enzymatic Methods

Examples

Calcium: Calcium + o-cresolphthalein complexone ——> coloured complex

Copper: Copper + bathocuproin disulfonate ——> coloured complex

Creatinine: Creatinine + picrate ——> coloured complex

Iron: Iron + bathophenanthroline disulfonate —> coloured complex

Phosphate (inorganic): Phosphate + ammonium molybdate + ammonium metavanadate ——> coloured complex

Enzymatic Methods

In enzymatic analysis (which is widely used in medical laboratories) the colour reaction is preceded by a reaction catalyzed by an enzyme. As the enzyme is specific to a particular substrate, more accurate results can be obtained. Enzymatic analysis is always carried out in a buffer solution at a specified temperature (usually 37°C) to provide the optimum conditions for the enzymes to act. Examples follow.

Examples

Cholesterol (CHOD-PAP method)

1. Cholesterol + oxygen —(enzyme cholesterol oxidase)—> cholestenone + hydrogen peroxide
2. Hydrogen peroxide + 4-aminophenazone + phenol —(enzyme peroxidase)—> coloured complex + water

Glucose (GOD-Perid method)

1. Glucose + oxygen + water —(enzyme glucose oxidase)—> gluconate + hydrogen peroxide
2. Hydrogen peroxide + ABTS —(enzyme peroxidase)—> coloured complex

In this case, both stages of the reaction are catalyzed by enzymes.

Triglycerides (GPO-PAP method)

1. Triglycerides + water —(enzyme esterase)—> glycerol + carboxylic acid
2. Glycerol + ATP —(enzyme glycerol kinase)—> glycerol-3-phosphate + ADP
3. Glycerol-3-phosphate + oxygen —(enzyme glycerol-3-phosphate oxidase) —> dihydroxyacetone phosphate + hydrogen peroxide

4. Hydrogen peroxide + 4-aminophenazone + 4-chlorophenol —(enzyme peroxidase)—> coloured complex

Urea

1. Urea + water —(enzyme urease)—> ammonium carbonate
2. Ammonium carbonate + phenol + hypochlorite ——> coloured complex

In this case, only the first stage of the reaction is catalyzed by an enzyme. The second stage is non-enzymatic.

Abbreviations

- CHOD = cholesterol oxidase
- GOD = glucose oxidase
- GPO = glycerol-3-phosphate oxidase
- PAP = phenol + aminophenazone (in some methods the phenol is replaced by 4-chlorophenol, which is less toxic)
- Perid = peroxidase

Ultraviolet Methods

In ultraviolet (UV) methods there is no visible colour change but the principle is exactly the same, i.e. the measurement of a change in the absorbance of the solution. UV methods usually measure the difference in absorbance at 340 nm wavelength between nicotinamide adenine dinucleotide (NAD) and its reduced form (NADH).

Examples

Pyruvate:

Pyruvate + NADH —(enzyme lactate dehydrogenase)—> L-lactate + NAD

Thin Layer Chromatography Analysis

Chromatography is the collective term for a set of laboratory techniques for the separation of mixtures. The mixture is dissolved in a fluid called the *mobile phase,* which carries it through a structure holding another material called the *stationary phase.* The various constituents of the mixture travel at different speeds, causing them to separate. The separation is based on differential partitioning between the mobile and stationary phases. Subtle differences in a compound's partition coefficient result in differential retention on the stationary phase and thus changing the separation.

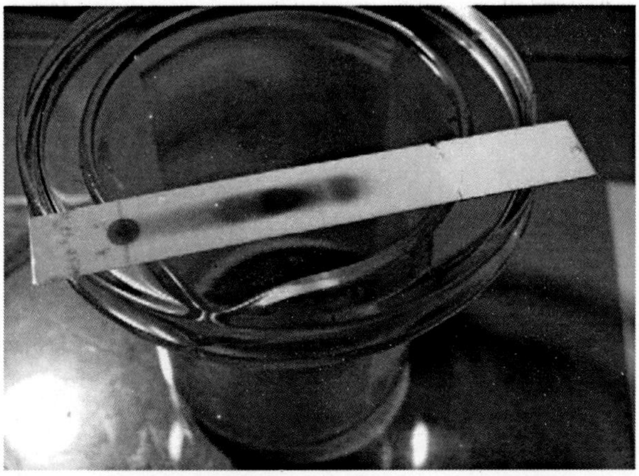

Figure: *Separation of black ink on a TLC plate*

Chromatography may be preparative or analytical. The purpose of preparative chromatography is to separate the components of a mixture for more advanced use (and is thus a form of purification). Analytical chromatography is done normally with smaller amounts of material and is for measuring the relative proportions of analytes in a mixture. The two are not mutually exclusive.

Thin-layer chromatography (TLC) is a chromatography technique used to separate non-volatile mixtures. Thin-layer chromatography is performed on a sheet of glass, plastic, or aluminium foil, which is coated with a thin layer of adsorbent material, usually silica gel, aluminium oxide, or cellulose. This layer of adsorbent is known as the stationary phase.

After the sample has been applied on the plate, a solvent or solvent mixture (known as the mobile phase) is drawn up the plate via capillary action. Because different analytes ascend the TLC plate at different rates, separation is achieved.

Thin-layer chromatography can be used to monitor the progress of a reaction, identify compounds present in a given mixture, and determine the purity of a substance. Specific examples of these applications include: analyzing ceramides and fatty acids, detection of pesticides or insecticides in food and water, analyzing the dye composition of fibres in forensics, assaying the radiochemical purity of radiopharmaceuticals, or identification of medicinal plants and their constituents

A number of enhancements can be made to the original method to automate the different steps, to increase the resolution achieved with TLC and to allow more accurate quantitative analysis. This method is referred to as HPTLC, or "high-performance TLC".

Plate Preparation

TLC plates are usually commercially available, with standard particle size ranges to improve reproducibility. They are prepared by mixing the adsorbent, such as silica gel, with a small amount of inert binder like calcium sulfate (gypsum) and water. This mixture is spread as a thick slurry on an unreactive carrier sheet, usually glass, thick aluminum foil, or plastic. The resultant plate is dried and *activated* by heating in an oven for thirty minutes at 110 °C. The thickness of the absorbent layer is typically around 0.1 – 0.25 mm for analytical purposes and around 0.5 – 2.0 mm for preparative TLC.

Technique

The process is similar to paper chromatography with the advantage of faster runs, better separations, and the choice between different stationary phases. Because of its simplicity and speed TLC is often used for monitoring chemical reactions and for the qualitative analysis of reaction products.

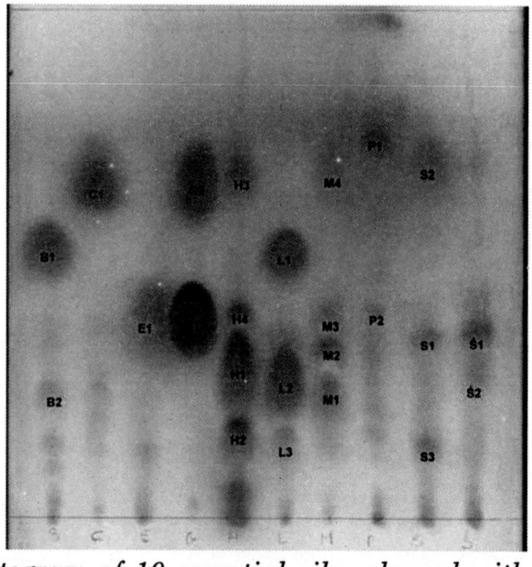

***Figure:** Chromatogram of 10 essential oils coloured with vanillin reagent.*

To run a thin layer chromatography, the following procedure is carried out:

- A small spot of solution containing the sample is applied to a plate, about 1.5 centimetres from the bottom edge. The solvent is allowed to completely evaporate off, otherwise a very poor or no separation will be achieved. If a non-volatile solvent was used to apply the sample, the plate needs to be dried in a vacuum chamber.

- A small amount of an appropriate solvent (eluent) is poured into a glass beaker or any other suitable transparent container (separation chamber) to a depth of less than 1 centimetre. A strip of filter paper (aka "wick") is put into the chamber so that its bottom touches the solvent and the paper lies on the chamber wall and reaches almost to the top of the container. The container is closed with a cover glass or any other lid and is left for a few minutes to let the solvent vapours ascend the filter paper and saturate the air in the chamber. (Failure to saturate the chamber will result in poor separation and non-reproducible results).
- The TLC plate is then placed in the chamber so that the spot(s) of the sample do not touch the surface of the eluent in the chamber, and the lid is closed. The solvent moves up the plate by capillary action, meets the sample mixture and carries it up the plate (elutes the sample). The plate should be removed from the chamber before the solvent front reaches the top of the stationary phase (continuation of the elution will give a misleading result) and dried.

Separation Process and Principle

Different compounds in the sample mixture travel at different rates due to the differences in their attraction to the stationary phase, and because of differences in solubility in the solvent. By changing the solvent, or perhaps using a mixture, the separation of components (measured by the R_f value) can be adjusted. Also, the separation achieved with a TLC plate can be used to estimate the separation of a flash chromatography column.

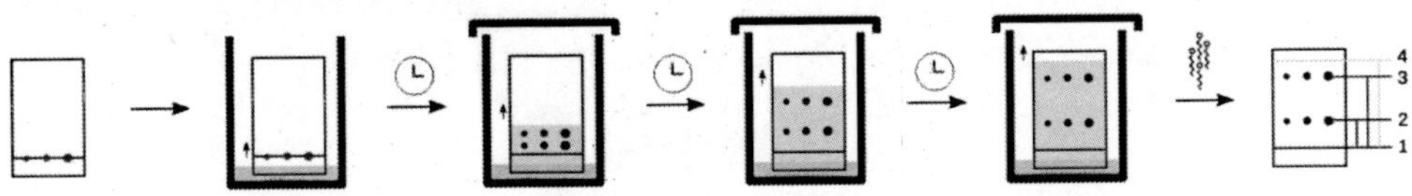

***Figure:** Development of a TLC plate, a purple spot separates into a red and blue spot.*

Separation of compounds is based on the competition of the solute and the mobile phase for binding places on the stationary phase. For instance, if normal phase silica gel is used as the stationary phase it can be considered polar. Given two compounds that differ in polarity, the more polar compound has a stronger interaction with the silica and is, therefore, more capable to dispel the mobile phase from the binding places. As a consequence, the less polar compound moves higher up the plate (resulting in a higher Rf value). If the mobile phase is changed to a more polar solvent or mixture of solvents, it is more capable of dispelling solutes from the silica binding places and all compounds on the TLC plate will move higher up the plate. It is commonly said that "strong" solvents (eluents) push the analyzed compounds up the plate, whereas "weak" eluents barely move them. The order of strength/ weakness depends on the coating (stationary phase) of the TLC plate. For silica gel coated TLC plates, the eluent strength increases in the following order: Perfluoroalkane (weakest), Hexane, Pentane, Carbon tetrachloride, Benzene/Toluene, Dichloromethane, Diethyl ether, Ethylacetate, Acetonitrile, Acetone, 2-Propanol/n-Butanol, Water, Methanol, Triethylamine, Acetic acid, Formic acid (strongest). For C18 coated plates the order is reverse.

This means that, if a mixture of ethyl acetate and hexane as the mobile phase is used, adding more ethyl acetate results in higher Rf values for all compounds on the TLC plate. *Changing the polarity of the mobile phase will normally not result in reversed order of running of the compounds on the TLC plate.* An eluotropic series can be used as a guide in selecting a mobile phase. If a reversed order of running of the compounds is desired, an apolar stationary phase should be used, such as C18-functionalized silica.

Analysis

As the chemicals being separated may be colourless, several methods exist to visualize the spots:

- fluorescent analytes like quinine may be detected under blacklight (366 nm)
- Often a small amount of a fluorescent compound, usually manganese-activated zinc silicate, is added to the adsorbent that allows the visualization of spots under UV-C light (254 nm). The adsorbent layer will thus fluoresce light-green by itself, but spots of analyte quench this fluorescence.
- Iodine vapours are a general unspecific colour reagent
- Specific colour reagents into which the TLC plate is dipped or which are sprayed onto the plate exist.
  - o Potassium permanganate - oxidation
  - o Bromine
- In the case of lipids, the chromatogram may be transferred to a PVDF membrane and then subjected to further analysis, for example mass spectrometry, a technique known as Far-Eastern blotting.

Once visible, the R_f value, or retardation factor, of each spot can be determined by dividing the distance the product travelled by the distance the solvent front travelled using the initial spotting site as reference. These values depend on the solvent used and the type of TLC plate and are not physical constants.

Applications

In organic chemistry, reactions are qualitatively monitored with TLC. Spots sampled with a capillary tube are placed on the plate: a spot of starting material, a spot from the reaction mixture, and a cross-spot with both. A small (3 by 7 cm) TLC plate takes a couple of minutes to run. The analysis is qualitative, and it will show if the starting material has disappeared, i.e. the reaction is complete, if any product has appeared, and how many products are generated (although this might be underestimated due to co-elution). Unfortunately, TLCs from low-temperature reactions may give misleading results, because the sample is warmed to room temperature in the capillary, which can alter the reaction—the warmed sample analyzed by TLC is not the same as what is in the low-temperature flask. One such reaction is the DIBALH reduction of ester to aldehyde.

As an example the chromatography of an extract of green leaves (for example spinach) in 7 stages of development. Carotene elutes quickly and is only visible until step 2. Chlorophyll A and B are halfway in the final step and lutein the first compound staining yellow.

In one study TLC has been applied in the screening of organic reactions, for example in the fine-tuning of BINAP synthesis from 2-naphthol. In this method, the alcohol and catalyst solution (for instance iron(III) chloride) are placed separately on the base line, then reacted, and then instantly analyzed.

Paper Chromatography Analysis

Chromatography is used to separate mixtures of substances into their components. All forms of chromatography work on the same principle.

They all have a *stationary phase* (a solid, or a liquid supported on a solid) and a *mobile phase* (a liquid or a gas). The mobile phase flows through the stationary phase and carries the components of the mixture with it. Different components travel at different rates. In paper chromatography, the stationary phase is a very uniform absorbent paper. The mobile phase is a suitable liquid solvent or mixture of solvents.

Producing a Paper Chromatogram

You probably used paper chromatography as one of the first things you ever did in chemistry to separate out mixtures of coloured dyes - for example, the dyes which make up a particular ink. That's an easy example to take, so let's start from there.

Suppose you have three blue pens and you want to find out which one was used to write a message. Samples of each ink are spotted on to a pencil line drawn on a sheet of chromatography paper. Some of the ink from the message is dissolved in the minimum possible amount of a suitable solvent, and that is also spotted onto the same line. In the diagram, the pens are labelled 1, 2 and 3, and the message ink as M.

Note: The chromatography paper will in fact be pure white - not pale grey.

The paper is suspended in a container with a shallow layer of a suitable solvent or mixture of solvents in it. It is important that the solvent level is below the line with the spots on it. The next diagram doesn't show details of how the paper is suspended because there are too many possible ways of doing it and it clutters the diagram. Sometimes the paper is just coiled into a loose cylinder and fastened with paper clips top and bottom. The cylinder then just stands in the bottom of the container.

The reason for covering the container is to make sure that the atmosphere in the beaker is saturated with solvent vapour. Saturating the atmosphere in the beaker with vapour stops the solvent from evaporating as it rises up the paper.

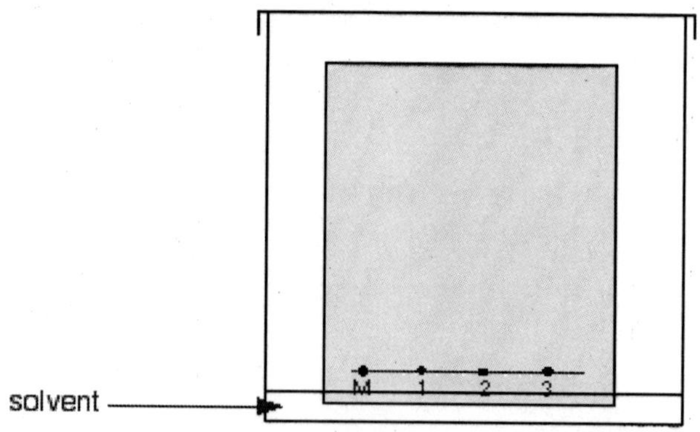

As the solvent slowly travels up the paper, the different components of the ink mixtures travel at different rates and the mixtures are separated into different coloured spots.

The diagram shows what the plate might look like after the solvent has moved almost to the top.

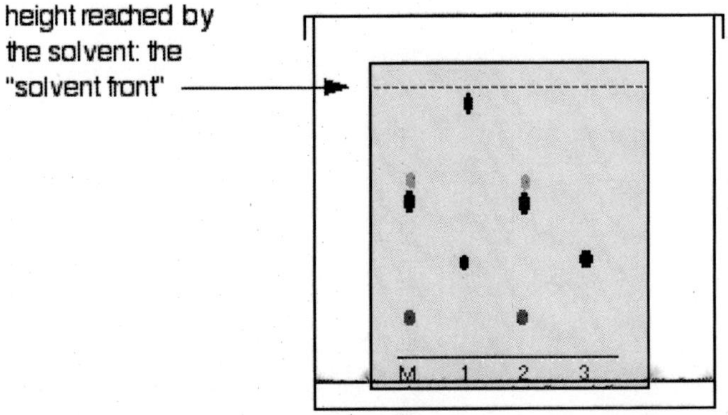

It is fairly easy to see from the final chromatogram that the pen that wrote the message contained the same dyes as pen 2. You can also see that pen 1 contains a mixture of two different blue dyes - one of which *might* be the same as the single dye in pen 3.

R_f Values

Some compounds in a mixture travel almost as far as the solvent does; some stay much closer to the base line. The distance travelled relative to the solvent is a constant for a particular compound as long as you keep everything else constant - the type of paper and the exact composition of the solvent, for example.

The distance travelled relative to the solvent is called the R_f value. For each compound it can be worked out using the formula:

For example, if one component of a mixture travelled 9.6 cm from the base line while the solvent had travelled 12.0 cm, then the R_f value for that component is:

In the example we looked at with the various pens, it wasn't necessary to measure R_f values because you are making a direct comparison just by looking at the chromatogram.

You are making the assumption that if you have two spots in the final chromatogram which are the same colour and have travelled the same distance up the paper, they are most likely the same compound. It isn't necessarily true of course - you could have two similarly coloured compounds with very similar R_f values. We'll look at how you can get around that problem further down the page.

What if the Substances you are Interested in are Colourless?

In some cases, it may be possible to make the spots visible by reacting them with something which produces a coloured product. A good example of this is in chromatograms produced from amino acid mixtures.

Suppose you had a mixture of amino acids and wanted to find out which particular amino acids the mixture contained. For simplicity we'll assume that you know the mixture can only possibly contain five of the common amino acids. A small drop of a solution of the mixture is placed on the base line of the paper, and similar small spots of the known amino acids are placed alongside it. The paper is then stood in a suitable solvent and left to develop as before. In the diagram, the mixture is M, and the known amino acids are labelled 1 to 5.

The position of the solvent front is marked in pencil and the chromatogram is allowed to dry and is then sprayed with a solution of *ninhydrin.* Ninhydrin reacts with amino acids to give coloured compounds, mainly brown or purple.

The left-hand diagram shows the paper after the solvent front has almost reached the top. The spots are still invisible. The second diagram shows what it might look like after spraying with ninhydrin.

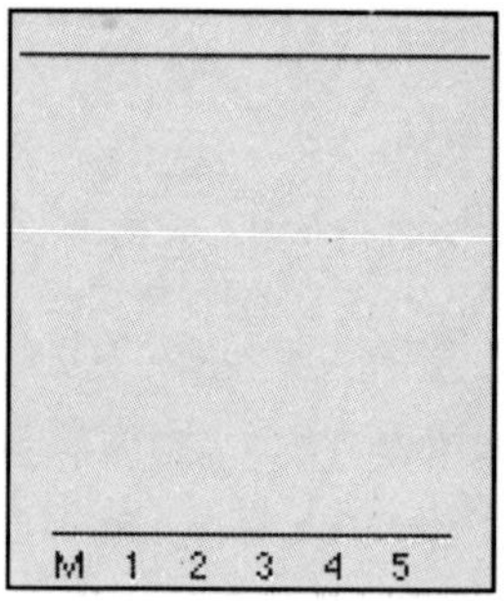

before spraying with ninhydrin

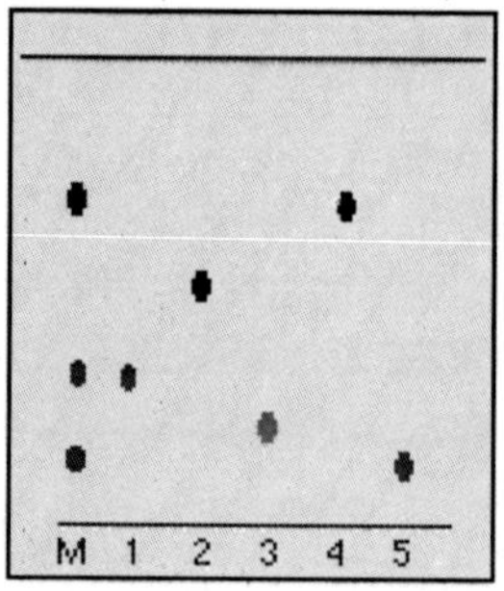

after spraying with ninhydrin

There is no need to measure the R_f values because you can easily compare the spots in the mixture with those of the known amino acids - both from their positions and their colours.

In this example, the mixture contains the amino acids labelled as 1, 4 and 5.

And what if the mixture contained amino acids other than the ones we have used for comparison? There would be spots in the mixture which didn't match those from the known amino acids. You would have to re-run the experiment using other amino acids for comparison.

Two Way Paper Chromatography

Two way paper chromatography gets around the problem of separating out substances which have very similar R_f values.

You can perfectly well do this with colourless compounds - but you have to use quite a lot of imagination in the explanation of what is going on!

This time a chromatogram is made starting from a single spot of mixture placed towards one end of the base line. It is stood in a solvent as before and left until the solvent front gets close to the top of the paper. In the diagram, the position of the solvent front is marked in pencil before the paper dries out. This is labelled as SF1 - the solvent front for the first solvent. We shall be using two different solvents.

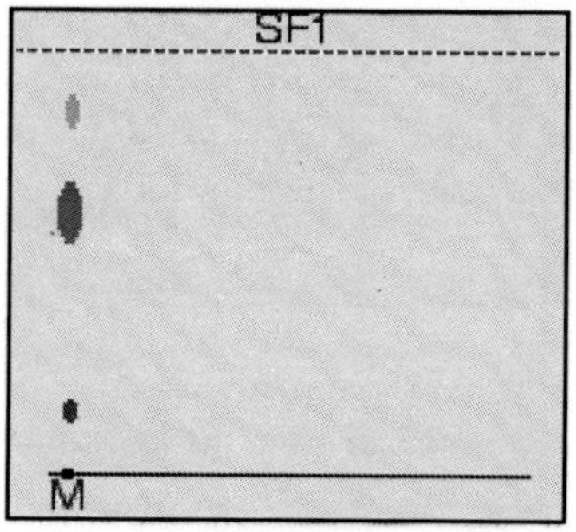

If you look closely, you may be able to see that the large central spot in the chromatogram is partly blue and partly green. Two dyes in the mixture have almost the same R_f values. They could equally well, of course, both have been the same colour - in which case you couldn't tell whether there was one or more dye present in that spot.

What you do now is to wait for the paper to dry out completely, and then rotate it through 90°, and develop the chromatogram again in a different solvent.

It is very unlikely that the two confusing spots will have the same R_f values in the second solvent as well as the first, and so the spots will move by a different amount.

The next diagram shows what might happen to the various spots on the original chromatogram. The position of the second solvent front is also marked.

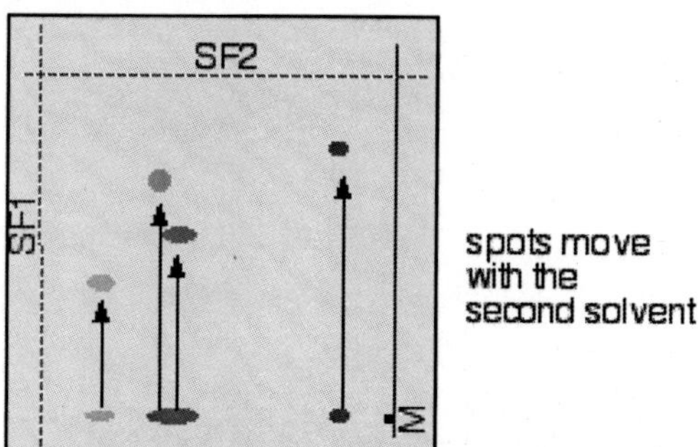

You wouldn't, of course, see these spots in both their original and final positions - they have moved! The final chromatogram would look like this:

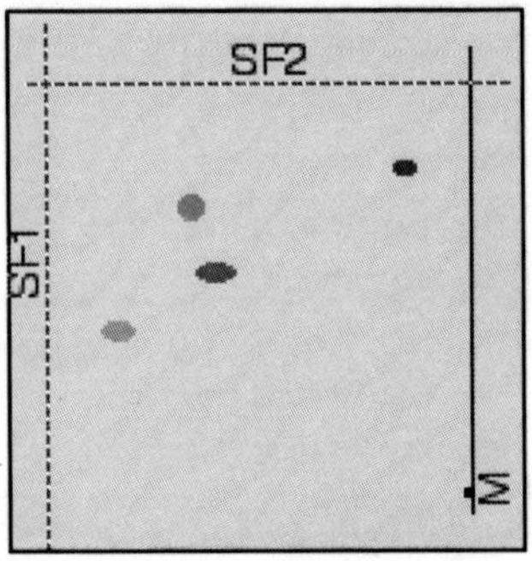

Two way chromatography has completely separated out the mixture into four distinct spots. If you want to identify the spots in the mixture, you obviously can't do it with comparison substances on the same chromatogram as we looked at earlier with the pens or amino acids examples. You would end up with a meaningless mess of spots. You can, though, work out the R_f values for each of the spots in both solvents, and then compare these with values that you have measured for known compounds under exactly the same conditions.

How does Paper Chromatography Work?

Although paper chromatography is simple to do, it is quite difficult to explain compared with thin layer chromatography. The explanation depends to some extent on what sort of solvent you are using, and many sources gloss over the problem completely. If you haven't already done so, it would be helpful if you could read the explanation for how thin layer chromatography works (link below). That will save me a lot of repetition, and I can concentrate on the problems.

The Essential Structure of Paper

Paper is made of cellulose fibres, and cellulose is a polymer of the simple sugar, glucose.

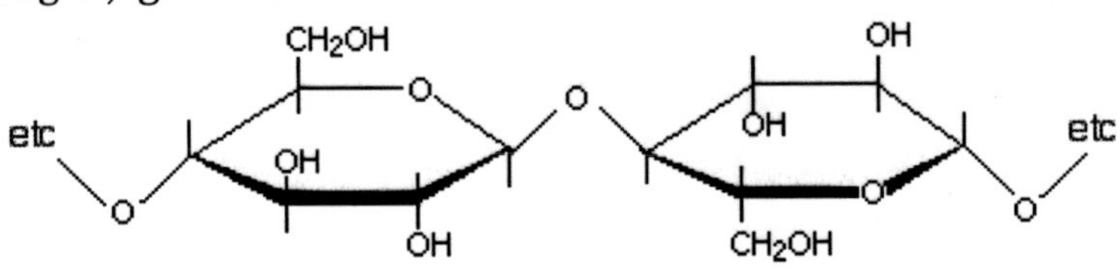

cellulose

The key point about cellulose is that the polymer chains have -OH groups sticking out all around them. To that extent, it presents the same sort of surface as silica gel or alumina in thin layer chromatography. It would be tempting to try to explain paper chromatography in terms of the way that different compounds are adsorbed to different extents on to the paper surface. In other words, it would be nice to be able to use the same explanation for both thin layer and paper chromatography.

Unfortunately, it is more complicated than that! The complication arises because the cellulose fibres attract water vapour from the atmosphere as well as any water that was present when the paper was made. You can therefore think of paper as being cellulose fibres with a very thin layer of water molecules bound to the surface. It is the interaction with this water which is the most important effect during paper chromatography.

Paper Chromatography using a Non-polar Solvent

Suppose you use a non-polar solvent such as hexane to develop your chromatogram. Non-polar molecules in the mixture that you are trying to separate will have little attraction for the water molecules attached to the cellulose, and so will spend most of their time dissolved in the moving solvent. Molecules like this will therefore travel a long way up the paper carried by the solvent. They will have relatively high R_f values. On the other hand, polar molecules *will* have a high attraction for the water molecules and much less for the non-polar solvent. They will therefore tend to dissolve in the thin layer of water around the cellulose fibres much more than in the moving solvent. Because they spend more time dissolved in the stationary phase and less time in the mobile phase, they aren't going to travel very fast up the paper. The tendency for a compound to divide its time between two immiscible solvents (solvents such as hexane and water which won't mix) is known as *partition*. Paper chromatography using a non-polar solvent is therefore a type of *partition chromatography*.

Paper Chromatography using a Water and other Polar Solvents

A moment's thought will tell you that partition can't be the explanation if you are using water as the solvent for your mixture. If you have water as the mobile phase and the water bound on to the cellulose as the stationary phase, there can't be any meaningful difference between the amount of time a substance spends in solution in either of them. All substances should be equally soluble (or equally insoluble) in both. And yet the first chromatograms that you made were probably of inks using water as your solvent.

If water works as the mobile phase as well being the stationary phase, there has to be some quite different mechanism at work - and that must be equally true for other polar solvents like the alcohols, for example. Partition only happens between solvents which don't mix with each other. Polar solvents like the small alcohols do mix with water.

2

Soil

Soil is the mixture of minerals, organic matter, gases, liquids and a myriad of organisms that can support plant life. It is a natural body that exists as part of the pedosphere and it performs four important functions: it is a medium for plant growth; it is a means of water storage, supply and purification; it is a modifier of the atmosphere; and it is a habitat for organisms that take part in decomposition and creation of a habitat for other organisms. Soil is considered the "skin of the earth" with interfaces between the lithosphere, hydrosphere, atmosphere, and biosphere. Soil consists of a solid phase (minerals & organic matter) as well as a porous phase that holds gases and water. Accordingly, soils are often treated as a three-state system.

Soil is the end product of the influence of the climate, relief (elevation, orientation, and slope of terrain), biotic activities (organisms), and parent materials (original minerals) interacting over periods of time. Soil continually undergoes development by way of numerous physical, chemical and biological processes, which include weathering with associated erosion. Most soils have a density between 1 and 2 g/cm^3. Little of the soil of planet Earth is older than the Pleistocene and none is older than the Cenozoic, although fossilized soils are preserved from as far back as the Archean.

Soil science has two main branches of study: Edaphology and Pedology. Pedology is focused on the formation, description (morphology), and classification of soils in their natural environment., whereas Edaphology is concerned with the influence of soils on organisms. In engineering terms, soil is referred to as regolith, or loose rock material that lies above the 'solid geology'. Soil is commonly referred to as "earth" or "dirt"; technically, the term "dirt" should be restricted to displaced soil. As soil resources serve as a basis for food security, the international community advocates for its sustainable and responsible use through different types of Soil Governance.

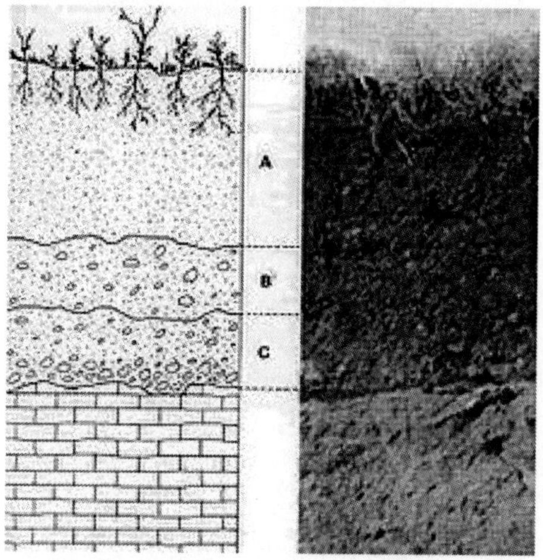

***Figure:** A represents soil; B represents laterite, a regolith; C represents saprolite, a less-weathered regolith; the bottom-most layer represents bedrock*

Organisms

Soil is the most abundant ecosystem on Earth, but the vast majority of organisms in soil are microbes, a great many of which have not been described. There may be a population limit of around one billion cells per gram of soil, but estimates of the number of species vary widely. Estimates range from over a million species per gram of soil to 50,000 per gram of soil. The total number of organisms and species can vary widely according to soil type, location, and depth.

Plants, animals, fungi, bacteria and humans affect soil formation. Animals, soil mesofauna and micro-organisms mix soils as they form burrows and pores, allowing moisture and gases to move about. In the same way, plant roots open channels in soils. Plants with deep taproots can penetrate many metres through the different soil layers to bring up nutrients from deeper in the profile. Plants with fibrous roots that spread out near the soil surface have roots that are easily decomposed, adding organic matter. Micro-organisms, including fungi and bacteria, effect chemical exchanges between roots and soil and act as a reserve of nutrients. Humans impact soil formation by removing vegetation cover with erosion as the result. Their tillage also mixes the different soil layers, restarting the soil formation process as less weathered material is mixed with the more developed upper layers. Earthworms, ants and termites mix the soil as they burrow, significantly affecting soil formation. Earthworms ingest soil particles and organic residues, enhancing the availability of plant nutrients in the material that passes through their bodies. They aerate and stir the soil and increase the stability of soil aggregates, thereby assuring ready infiltration of water. As they build mounds, some organisms might transport soil materials from one horizon to another. In general, the mixing activities of animals, sometimes called pedoturbation, tends to undo or counteract the tendency of other soil-forming processes that create distinct horizons. Termites and ants may also retard soil profile development by denuding large areas of soil around their nests, leading to increased loss of soil by erosion. Large animals such as gophers, moles, and prairie dogs bore into the lower soil horizons, bringing materials to the surface. Their tunnels are often open to the surface, encouraging the movement of water and air into the subsurface layers. In localized areas, they enhance mixing of the lower and upper horizons by creating, and later refilling, underground tunnels. Old animal burrows in the lower horizons often become filled with soil material from the overlying A horizon, creating profile features known as crotovinas.

Vegetation impacts soils in numerous ways. It can prevent erosion caused by excessive rain that might result from surface runoff. Plants shade soils, keeping them cooler and slow evaporation of soil moisture, or conversely, by way of transpiration, plants can cause soils to lose moisture. Plants can form new chemicals that can break down minerals and improve the soil structure. The type and amount of vegetation depends on climate, topography, soil characteristics, and biological factors. Soil factors such as density, depth, chemistry, pH, temperature and moisture greatly affect the type of plants that can grow in a given location. Dead plants and fallen leaves and stems begin their decomposition on the surface. There, organisms feed on them and mix the organic material with the upper soil layers; these added organic compounds become part of the soil formation process.

Human activities widely influence soil formation. For example, it is believed that Native Americans regularly set fires to maintain several large areas of prairie grasslands in Indiana and Michigan. In more recent times, human destruction of natural vegetation and subsequent tillage of the soil for crop production has abruptly modified soil formation. Likewise, irrigating an arid region of soil drastically influences the soil-forming factors, as does adding fertilizer and lime to soils of low fertility.

Soil Atmosphere

The atmosphere of soil is radically different from the atmosphere above. The consumption of oxygen, by microbes and plant roots and their release of carbon dioxide, decrease oxygen and increase carbon dioxide concentration. Atmospheric CO_2 concentration is 0.04%, but in the soil pore space it may range from 10 to 100 times that level. At extreme levels CO_2 is toxic. In addition, the soil voids are saturated with water vapour. Adequate porosity is necessary not just to allow the penetration of water but also to allow gases to diffuse in and out. Movement of gases is by diffusion from high concentrations to lower. Oxygen diffuses in and is consumed and excess levels of carbon dioxide, diffuse out with other gases as well as water. Soil texture and structure strongly affect soil porosity and gas diffusion. Platy and compacted soils impede gas flow, and a deficiency of oxygen may encourage anaerobic bacteria to reduce nitrate to the gases N_2, N_2O, and NO, which are then lost to the atmosphere. Aerated soil is also a net sink of methane CH_4 but a net producer of greenhouse gases when soils are depleted of oxygen and subject to elevated temperatures.

Colloidal and Chemical Properties

The chemistry of soil determines the availability of nutrients, the health of microbial populations, and its physical properties. In addition, soil chemistry also determines its corrosivity, stability, and ability to absorb pollutants and to filter water. It is the surface chemistry of clay and humus colloids that determines soil's chemical properties. "A colloid is a small, insoluble, nondiffusible particle larger than a molecule but small enough to remain suspended in a fluid medium without settling. Most soils contain organic colloidal particles called humus as well as the inorganic colloidal particles of clays." The very high specific surface area of colloids and their net negative charges, gives soil its great ability to hold and release cations in what is referred to as cation exchange. Cation-exchange capacity (CEC) is the amount of exchangeable cations per unit weight of dry soil and is expressed in terms of milliequivalents of hydrogen ion per 100 grams of soil.

Soil Quality

Soil quality is a measure of the condition of soil relative to the requirements of one or more biotic species and or to any human need or purpose. According to the United States Department of Agriculture Natural Resources Conservation Service, "Soil quality is the capacity of a specific kind of soil to function, within natural or managed ecosystem boundaries, to sustain plant and animal productivity, maintain or enhance water and air quality, and support human health and habitation." The European Commission's Joint Research Centre proposed a definition, stating that "Soil quality is an account of the soil's ability to provide ecosystem and social services through its capacities to perform its functions under changing conditions."

Soil quality reflects how well a soil performs the functions of maintaining biodiversity and productivity, partitioning water and solute flow, filtering and buffering, nutrient cycling, and providing support for plants and other structures. Soil management has a major impact on soil quality.

Amino Acids in Soils

Amino acids are biologically important organic compounds composed of amine ($-NH_2$) and carboxylic acid (-COOH) functional groups, along with a side-chain specific to each amino acid. The key elements of an amino acid are carbon, hydrogen, oxygen, and nitrogen, though other elements are found in the side-chains of certain amino acids. About 500 amino acids are known and can be classified in many ways. They can be classified according to the core structural functional groups' locations as alpha- (α-), beta- (β-), gamma- (γ-) or delta- (δ-) amino acids; other categories relate to polarity, pH level, and side-chain group type (aliphatic, acyclic, aromatic, containing hydroxyl or sulfur, etc.). In the form of proteins, amino acids comprise the second-largest component (water is the largest) of human muscles, cells and other tissues. Outside proteins, amino acids perform critical roles in processes such as neurotransmitter transport and biosynthesis.

Amino acids having both the amine and the carboxylic acid groups attached to the first (alpha-) carbon atom have particular importance in biochemistry. They are known as 2-, alpha-, or α-amino acids (generic formula $H_2NCHRCOOH$ in most cases where R is an organic substituent known as a "side-chain"); often the term "amino acid" is used to refer specifically to these. They include the 23 proteinogenic ("protein-building") amino acids, which combine into peptide chains ("polypeptides") to form the building-blocks of a vast array of proteins. These are all L-stereoisomers ("left-handed" isomers), although a few D-amino acids ("right-handed") occur in bacterial envelopes and some antibiotics. Twenty of the proteinogenic amino acids are encoded directly by triplet codons in the genetic code and are known as "standard" amino acids. The other two ("non-standard" or "non-canonical") are pyrrolysine (found in methanogenic organisms and other eukaryotes) and selenocysteine (present in many noneukaryotes as well as most eukaryotes). For example, 25 human proteins include selenocysteine (Sec) in their primary structure, and the structurally characterized enzymes (selenoenzymes) employ Sec as the catalytic moiety in their active sites. Pyrrolysine and selenocysteine are encoded via variant codons; for example, selenocysteine is encoded by stop codon and SECIS element. Codon–tRNA combinations not found in nature can also be used to "expand" the genetic code and create novel proteins known as alloproteins incorporating non-proteinogenic amino acids.

Many important proteinogenic and non-proteinogenic amino acids also play critical non-protein roles within the body. For example, in the human brain, glutamate (standard glutamic acid) and gamma-amino-butyric acid ("GABA", non-standard gamma-amino acid) are, respectively, the main excitatory and inhibitory neurotransmitters; hydroxyproline (a major component of the connective tissue collagen) is synthesised from proline; the standard amino acid glycine is used to synthesise porphyrins used in red blood cells; and the non-standard carnitine is used in lipid transport. Nine proteinogenic amino acids are called "essential" for humans because they cannot be created from other compounds by the human body and, so, must be taken in as food. Others may be conditionally essential for certain ages or medical conditions. Essential amino acids may also differ between species.

Because of their biological significance, amino acids are important in nutrition and are commonly used in nutritional supplements, fertilizers, and food technology. Industrial uses include the production of drugs, biodegradable plastics, and chiral catalysts.

The first few amino acids were discovered in the early 19th century. In 1806, French chemists Louis-Nicolas Vauquelin and Pierre Jean Robiquet isolated a compound in asparagus that was subsequently named asparagine, the first amino acid to be discovered. Cystine was discovered in 1810, although its monomer, cysteine, remained undiscovered until 1884. Glycine and leucine were discovered in 1820. Usage of the term *amino acid* in the English language is from 1898. Proteins were found to yield amino acids after enzymatic digestion or acid hydrolysis. In 1902, Emil Fischer and Franz Hofmeister proposed that proteins are the result of the formation of bonds between the amino group of one amino acid with the carboxyl group of another, in a linear structure that Fischer termed peptide.

General Structure

In the structure shown at the top of the page, R represents a side-chain specific to each amino acid. The carbon atom next to the carboxyl group is called the α–carbon and amino acids with a side-chain bonded to this carbon are referred to as *alpha amino acids*. These are the most common form found in nature. In the alpha amino acids, the α–carbon is a chiral carbon atom, with the exception of glycine.

In amino acids that have a carbon chain attached to the α–carbon (such as lysine, shown to the right) the carbons are labeled in order as α, β, γ, δ, and so on. In some amino acids, the amine group is attached to the β or γ-carbon, and these are therefore referred to as *beta* or *gamma amino acids*.

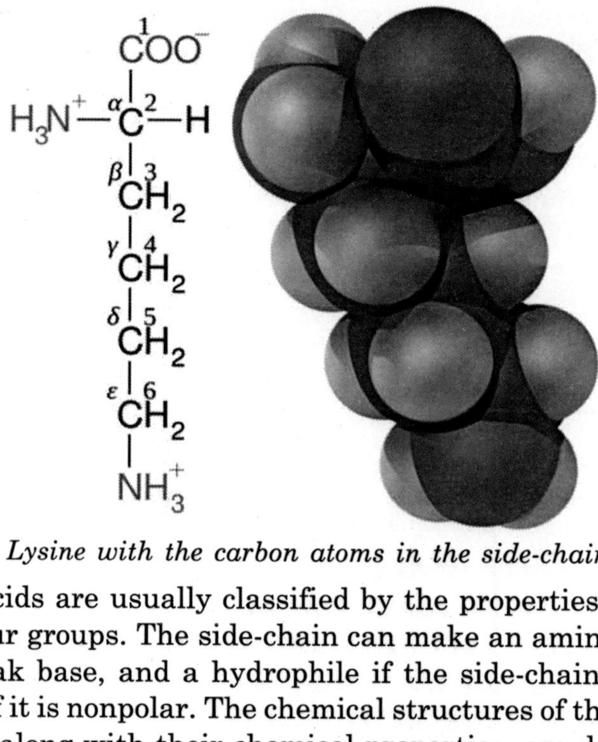

Figure: *Lysine with the carbon atoms in the side-chain labeled*

Amino acids are usually classified by the properties of their side-chain into four groups. The side-chain can make an amino acid a weak acid or a weak base, and a hydrophile if the side-chain is polar or a hydrophobe if it is nonpolar. The chemical structures of the 22 standard amino acids, along with their chemical properties, are described more fully in the article on these proteinogenic amino acids.

The phrase "branched-chain amino acids" or BCAA refers to the amino acids having aliphatic side-chains that are non-linear; these are leucine, isoleucine, and valine. Proline is the only proteinogenic amino acid whose side-group links to the α-amino group and, thus, is also the only proteinogenic amino acid containing a secondary amine at this position. In chemical terms, proline is, therefore, an imino acid, since it lacks a primary amino group, although it is still classed as an amino acid in the current biochemical nomenclature, and may also be called an "N-alkylated alpha-amino acid".

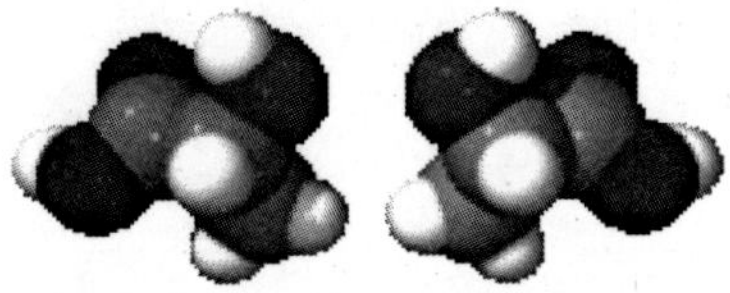

Figure: *The two enantiomers of alanine, D-Alanine and L-Alanine*

Isomerism

Of the standard α-amino acids, all but glycine can exist in either of two enantiomers, called L or D amino acids, which are mirror images of each other. While L-amino acids represent all of the amino acids found in proteins during translation in the ribosome, D-amino acids are found in some proteins produced by enzyme posttranslational modifications after translation and translocation to the endoplasmic reticulum, as in exotic sea-dwelling organisms such as cone snails. They are also abundant components of the peptidoglycan cell walls of bacteria, and D-serine may act as a neurotransmitter in the brain. D-amino acids are used in racemic crystallography to create centrosymmetric crystals, which (depending on the protein) may allow for easier and more robust protein structure determination.

The L and D convention for amino acid configuration refers not to the optical activity of the amino acid itself but rather to the optical activity of the isomer of glyceraldehyde from which that amino acid can, in theory, be synthesized (D-glyceraldehyde is dextrorotatory; L-glyceraldehyde is levorotatory). In alternative fashion, the *(S)* and *(R)* designators are used to indicate the absolute stereochemistry. Almost all of the amino acids in proteins are *(S)* at the α carbon, with cysteine being *(R)* and glycine non-chiral. Cysteine is unusual since it has a sulfur atom at the second position in its side-chain, which has a larger atomic mass than the groups attached to the first carbon, which is attached to the α-carbon in the other standard amino acids, thus the *(R)* instead of *(S)*.

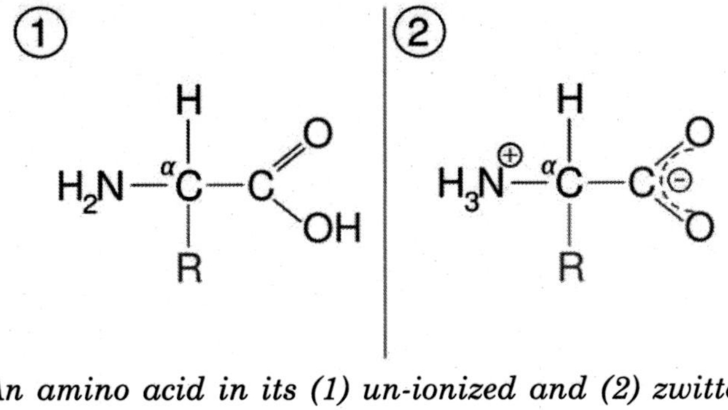

Figure: *An amino acid in its (1) un-ionized and (2) zwitterionic forms*

Zwitterions

The amine and carboxylic acid functional groups found in amino acids allow them to have amphiprotic properties. Carboxylic acid groups ($-CO_2H$) can be deprotonated to become negative carboxylates ($-CO_2^-$), and α-amino groups (NH_2-) can be protonated to become positive α-ammonium groups ($^+NH_3-$). At pH values greater than the pKa of the carboxylic acid group, the negative carboxylate ion predominates. At pH values lower than the pKa of the α-ammonium group (mean for the 20 common α-amino acids is about 9.4), the nitrogen is predominantly protonated as a positively charged α-ammonium group. Thus, at pH between 2.2 and 9.4, the predominant form adopted by α-amino acids contains a negative carboxylate and a positive α-ammonium group, as shown in structure (2) on the right, so has net zero charge. This molecular state is known as a zwitterion, from the German Zwitter meaning *hermaphrodite* or *hybrid.* Below pH 2.2, the predominant form will have a neutral carboxylic acid group and a positive α-ammonium ion (net charge +1), and above pH 9.4, a negative carboxylate and neutral α-amino group (net charge –1). The fully neutral form (structure (1) on the right) is a very minor species in aqueous solution throughout the pH range (less than 1 part in 10^7). Amino acids exist as zwitterions also in the solid phase, and crystallize with salt-like properties unlike typical organic acids or amines.

Isoelectric Point

At pH values between the two pKa values, the zwitterion predominates, but coexists in dynamic equilibrium with small amounts of net negative and net positive ions. At the exact midpoint between the two pKa values, the trace amount of net negative and trace of net positive ions exactly balance, so that average net charge of all forms present is zero. This pH is known as the isoelectric point pI, so pI = $\frac{1}{2}(pKa_1 + pKa_2)$. The individual amino acids all have slightly different pKa values, so have different isoelectric points. For amino acids with charged side-chains, the pKa of the side-chain is involved. Thus for Asp, Glu with negative side-chains, pI = $\frac{1}{2}(pKa_1 + pKa_R)$, where pKa_R is the side-chain pKa. Cysteine also has potentially negative side-chain with $pKa_R = 8.14$, so pI should be calculated as for Asp and Glu, even though the side-chain is not significantly charged at neutral pH. For His, Lys, and Arg with positive side-chains, pI = $\frac{1}{2}(pKa_R + pKa_2)$. Amino acids have zero mobility in electrophoresis at their isoelectric point, although this behaviour is more usually exploited for peptides and proteins than single amino acids. Zwitterions have minimum solubility at their isolectric point and some amino acids (in particular, with non-polar side-chains) can be isolated by precipitation from water by adjusting the pH to the required isoelectric point.

Occurrence and Functions in Biochemistry

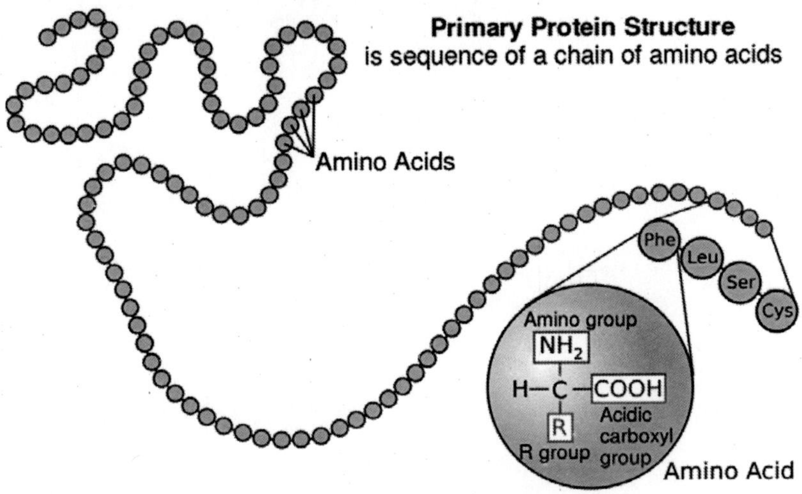

***Figure:** A polypeptide is an unbranched chain of amino acids.*

Essential Amino Acids

Amino acids are the structural units (monomers) that make up proteins. They join together to form short polymer chains called peptides or longer chains called either polypeptides or proteins. These polymers are linear and unbranched, with each amino acid within the chain attached to two neighbouring amino acids. The process of making proteins is called *translation* and involves the step-by-step addition of amino acids to a growing protein chain by a ribozyme that is called a ribosome. The order in which the amino acids are added is read through the genetic code from an mRNA template, which is a RNA copy of one of the organism's genes.

Twenty-two amino acids are naturally incorporated into polypeptides and are called proteinogenic or natural amino acids. Of these, 20 are encoded by the universal genetic code. The remaining 2, selenocysteine and pyrrolysine, are incorporated into proteins by unique synthetic mechanisms. Selenocysteine is incorporated when the mRNA being translated includes a SECIS element, which causes the UGA codon to encode selenocysteine instead of a stop codon. Pyrrolysine is used by some methanogenic archaea in enzymes that they use to produce methane. It is coded for with the codon UAG, which is normally a stop codon in other organisms. This UAG codon is followed by a PYLIS downstream sequence.

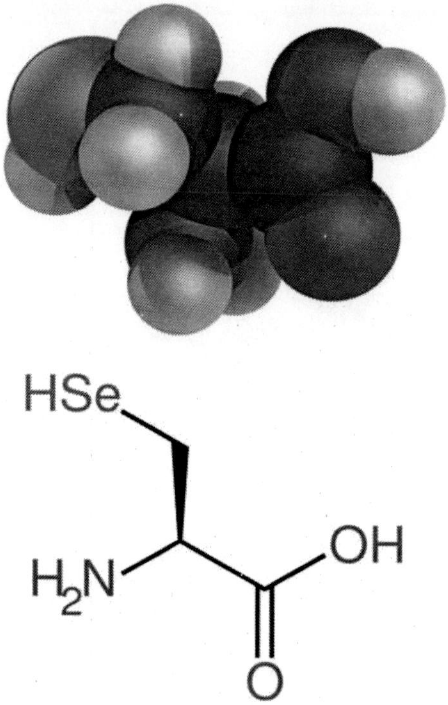

Figure: *The amino acid selenocysteine*

Non-standard Amino Acids

Aside from the 23 proteinogenic amino acids, there are many other amino acids that are called *non-proteinogenic* or *non-standard* (although three of the proteinogenic amino acids are also called *non-standard*, not being encoded simply by the universal genetic code but requiring additional contextual information). Those either are not found in proteins (for example carnitine, GABA) or are not produced directly and in isolation by standard cellular machinery (for example, hydroxyproline and selenomethionine).

Non-proteinogenic amino acids that are found in proteins are formed by post-translational modification, which is modification after translation during protein synthesis. These modifications are often essential for the function or regulation of a protein; for example, the carboxylation of glutamate allows for better binding of calcium cations, and the hydroxylation of proline is critical for maintaining connective tissues. Another example is the formation of hypusine in the translation initiation factor EIF5A, through modification of a lysine residue. Such modifications can also determine the localization of the protein, e.g., the addition of long hydrophobic groups can cause a protein to bind to a phospholipid membrane.

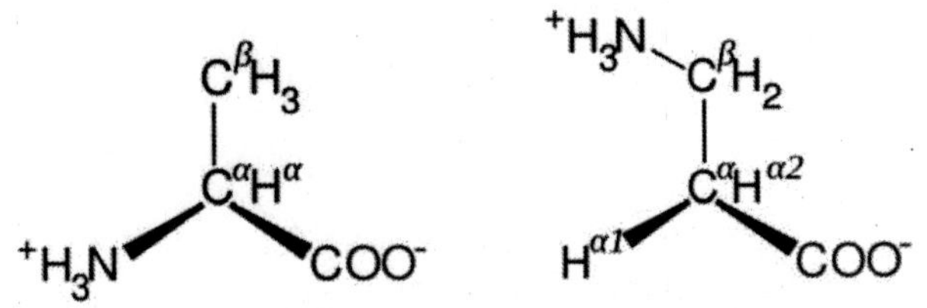

***Figure:** β-alanine and its α-alanine isomer*

Some nonstandard amino acids are not found in proteins. Examples include lanthionine, 2-aminoisobutyric acid, dehydroalanine, and the neurotransmitter gamma-aminobutyric acid. Nonstandard amino acids often occur as intermediates in the metabolic pathways for standard amino acids – for example, ornithine and citrulline occur in the urea cycle, part of amino acid catabolism. A rare exception to the dominance of α-amino acids in biology is the β-amino acid beta alanine (3-aminopropanoic acid), which is used in plants and microorganisms in the synthesis of pantothenic acid (vitamin B_5), a component of coenzyme A.

Classification

Although there are many ways to classify amino acids, these molecules can be assorted into six main groups, on the basis of their structure and the general chemical characteristics of their R groups.

Class	*Name of the amino acids*
Aliphatic	Glycine, Alanine, Valine, Leucine, Isoleucine
Hydroxyl or Sulfur/Selenium-containing	Serine, Cysteine, Selenocysteine, Threonine, Methionine
Cyclic	Proline
Aromatic	Phenylalanine, Tyrosine, Tryptophan
Basic	Histidine, Lysine, Arginine
Acidic and their Amide	Aspartate, Glutamate, Asparagine, Glutamine

Organic Form of Soil Nitrogen

Over 90% of the nitrogen N in the surface layer of most soils occurs in organic forms, with most of the remainder being present as NH_4^- whichis held within the lattice structures of clayminerals. The surface layerof most cultivated soils contains between 0.06 and 0.3% N. Peat soils have high N contents to 3.5%.

Plant remains and other debris contribute nitrogen N in the form of:

Amino Acids

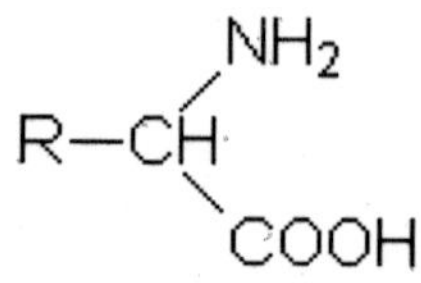

The general formula of amino acids.

(Stevenson 1982)

Figure: *Amino acids exist in soil in several different forms, like:*

1. As free amino acids
   - o in the soil solution
   - o in soil micropores
2. As amino acids, peptides or proteins bound to clay minerals
   - o on external surfaces
   - o on internal surfaces
3. As amino acids, peptides or proteins bound to humic colloids
   - o H-bonding and van der Waals' forces
   - o in covalent linkage as quinoid-amino acid complexes
4. As mucoproteins
5. As a muramic acid

Amino acids, being readily decomposed by microorganisms, have only anephemeral existence in soil. Thus the amounts present in the soil solutionat any one time represent a balance between synthesis and destruction by microorganisms.

The free amino acids content of the soil is strongly influenced by weatherconditions, moisture status of the soil, type of plant and stage of growth, additions of organic residues, and cultural conditions.

Amino Sugars

Amino sugars occur as structural components of a broad group of substances, the mucopolysaccharides and they have been found in combination with mucopeptides and mucoproteins. Some of the amino sugar material in soil may exist in the form of an alkali-insoluble polysaccharide referred to as chitin.

Generally the amino sugars in soil are of microbial origin. From 5 to 10% of the N in the surface layer of most soils can be accounted for in N-containing carbohydrates or amino sugars.

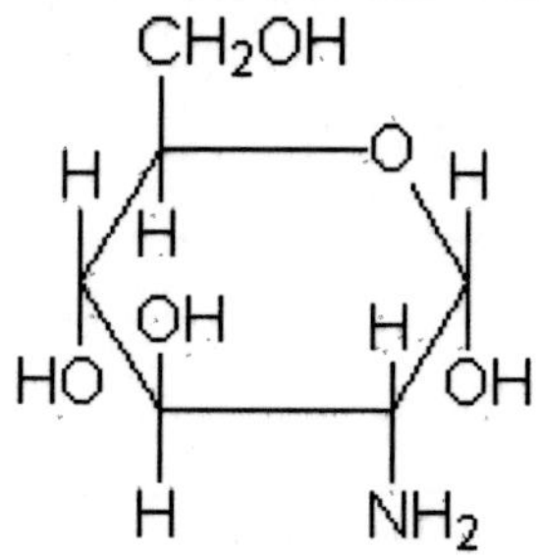

D - Glucosamine
(Stevenson 1982)

Nucleic Acids

Nucleic acids, which occur in the cells of all living organisms, consist of individual mononucleotide units (base-sugar-phosphate) joined by a phosphoricacid ester linkage through the sugar. Two types: ribonucleic acid (RNA) anddeoxyribonucleic acid (DNA). They have pentose sugar (ribose or deoxyribose), the purine: adenine, guanine and the pyrimidine: cytosine, thymine. RNA contains also the uracil.

The N in purine and pyrimidine bases is usually considered to account forless than 1% of the total soil N.

Small amounts of N are extracted from soil in the form of glycerophosphatides, amines, vitamins, pesticide and pesticide degradation products.

Nitrogen Transformation

A key feature of the internal cycle is the biological turnover of N between mineral and organic forms through the opposing processes of mineralization and immobilization. The latter leads to incorporation of N into microbial tissues. Whereas much of this newly immobilized N is recycled through mineralization, some is converted to stable humus forms.

The overall reaction leading to incorporation of inorganic forms of N in to stable humus forms is depicted on the picture.

Thus the decay of plant and animal residues by microorganisms results in the formation of mineral forms of N (NH_4^+ and NO_3^-) and assimilation of part of the C into microbial tissue (reaction A). Part of the native humus undergoes a similar fate (reaction B). Subsequent turnover through mineralization-immobilization leads to incorporation of N into stable humus forms (reaction C). Stabilization of N may also occur through the reaction of partial decay products of lignin with nitrogenous constituents (raection D).

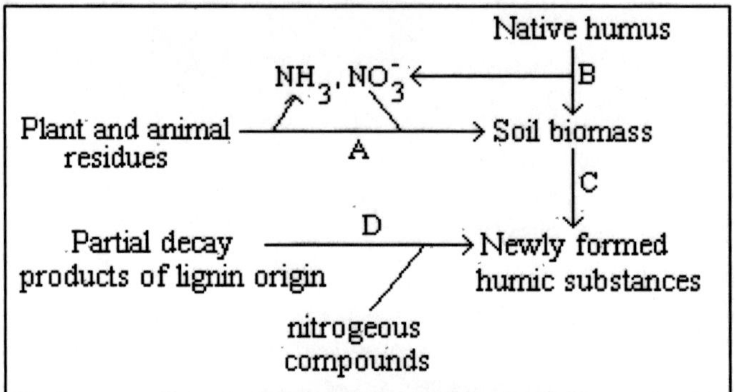

The incorporation of organic forms of N into stable humus forms. (Stevenson 1982)

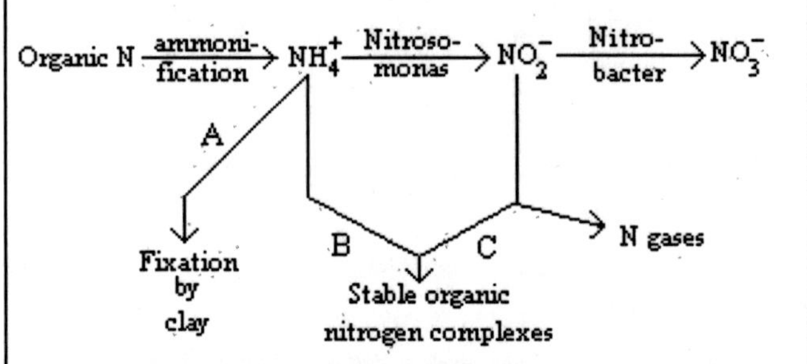

Nonbiological transformations of NH_3 and NO_2^- in soil. (Stevenson 1982)

Except under unusual circumstances, both mineralization and immobilization always function in soil, but in opposite direction.

Chemical Reaction of Ammonia and Nitrite with Organic Matter

The fate of mineral forms of N in soil is determined to some extent by nonbiological reactions involving NH_4^+, NH3 and NO_2^- as depicted in fig.

In addition to NH_4^+ fixation by clay minerals (reaction A), NH_3 and NO_2^- react chemically with organic matter to form stable organic N complexes (reaction B and C). The chemical interaction of NO_2^- with organic matter may lead to the generation of N gases. Although both types of reactions can proceed over a wide pH range, fixation of NH_3^+ is favoured by a high pH (>7.0). In contrast, NO_2^- -organic matter interactions occur most readily under highly acidic conditions (pH of 5.0 to 5.5 or below).

Stability of Soil Organic Nitrogen

1. Proteinaceous constituents are stabilized through their reaction with other organic constituents, such as lignins, tanins, quinons.
2. Biologically resistant complexes are formed in soil by chemical reactions involving NH_3^+ or NO_2^- with lignins or humic substances.
3. Adsorption of organic N compounds by clay minerals (pariculary montmorillinitic types) protects the molecule from decomposition.
4. Complexes formed between organic N compounds and polyvalent cations, such as Fe, are biologically stable.
5. Some of the organic N occurs in small pores or voids and is physically inaccessible to microorganisms.

C/N Ratio

For surface soils, and for the top layer of lake and marine sediments, the ratio generally falls within well-defined limits, usually from about 10 to 12. In most soils, the C/N ratio decreases with increasing depth, often attaining values less than 5.0. Native humus would be expected to have a lower C/N ratio than most undecayed plant residues for following reasons. The decay of organic residues by soilorganisms leads to incorporation of part of the C into microbial tissue with the remainder being liberated as CO_2. As a general rule, about one-third of the applied C in fresh residues will remain in the soil after the first few months of decomposition. The decay process is accompanied by conversion of organic form of N to NH_3 and NO_3^- and soil microorganisms utilize part of this N for synthesis of new cells. The gradual transformation of plantraw material into stable organic matter (humus) leads to the establishment of reasonably consistent relationship between C and N. Other factors which may be involved in narrowing of the C/N ratio include chemical fixation of NH_3 or amines by lignin like substances.

The C/N ratio of virigin soils formed under grass vegetation is normally lower than for soils formed under forest vegetation, and for the latter, the C/N ratio of the humus layers is usually higher than for the mineral soil proper. Also the C/N ratio of a well-decomposed muck soil is lower than for a fibrous peat.

As a general rule it can be said that conditions which encourage decomposition of organic matter result in narrowing of the C/N ratio. The ratio nearly always narrows sharply with depth in the profile; for certain subsurface soils C/Nratios lower than 5 are not uncommon.

Effect of Amino Acids on Plants

Agriculture production is a very intensive business and is related to better quality and better yield leading to better profitability.

Every Farmers dreams to achieve this goal. However to achieve this goal with advancement of technology, use of fertilizer and pesticides is not adequate. Now is the time to look at Bioenergetics and Biochemical aspects of plants, to achieve the goals of Farmerss.

Every plant like any organism needs certain components for growth over and above soil, sun, rain and air. The basic component of living cells is Proteins, with building block material, Amino Acids. Proteins are formed by sequence of Amino Acids.

Plants synthesize Amino Acids from the Primary elements, the Carbon and Oxygen obtained from air, Hydrogen from water in the soil, forming Carbon Hydrate by means of photosynthesis and combining it with the Nitrogen which the plants obtain from the soil, leading to synthesis of amino acids, by collateral metabolic pathways. Only L-Amino Acids are part of these Proteins and have metabolic activity.

The requirement of amino acids in essential quantities is well known as a means to increase yield and overall quality of crops.

The application of amino acids for foliar use is based on its requirement by plants in general and at critical stages of growth in particular. Plants absorb Amino Acids through Stomas and is proportionate to environment temperature.

Amino Acids are fundamental ingredients in the process of Protein Synthesis. About 20 important Amino Acids are involved in the process of each function. Studies have proved that Amino Acids can directly or indirectly influence the physiological activities of the plant.

Amino Acids are also supplied to plant by incorporating them into the soil. It helps in improving the microflora of the soil thereby facilitating the assimilation of nutrients.

Foliar Nutrition in the form of Protein Hydrolysate (Known as Amino Acids Liquid) and foliar spray provide readymade building blocks for Protein synthesis.

Protein Synthesis

Proteins have a structural function, metabolic function (enzymes), a transport function and a stock of Amino Acids function.

Only L - Amino Acids are assimilated by plants. D - Amino Acids are not recognised by the enzymatic locus and therefore can not participate in protein synthesis.

Hence Amino Acids obtained by organic synthesis are not well assimilated.

Soil Enzymes

Enzymes are large biological molecules responsible for the thousands of metabolic processes that sustain life. They are highly selective catalysts, greatly accelerating both the rate and specificity of metabolic reactions, from the digestion of food to the synthesis of DNA. Most enzymes are proteins, although some catalytic RNA molecules have been identified. Enzymes adopt a specific three-dimensional structure, and may employ organic (e.g. biotin) and inorganic (e.g. magnesium ion) cofactors to assist in catalysis.

In enzymatic reactions, the molecules at the beginning of the process, called substrates, are converted into different molecules, called products. Almost all chemical reactions in a biological cell need enzymes in order to occur at rates sufficient for life. Since enzymes are selective for their substrates and speed up only a few reactions from among many possibilities, the set of enzymes made in a cell determines which metabolic pathways occur in that cell.

Like all catalysts, enzymes work by lowering the activation energy ($E_a^{\ddagger}$) for a reaction, thus dramatically increasing the rate of the reaction. As a result, products are formed faster and reactions reach their equilibrium state more rapidly. Most enzyme reaction rates are millions of times faster than those of comparable un-catalyzed reactions. As with all catalysts, enzymes are not consumed by the reactions they catalyze, nor do they alter the equilibrium of these reactions. However, enzymes do differ from most other catalysts in that they are highly specific for their substrates. Enzymes are known to catalyze about 4,000 biochemical reactions. A few RNA molecules called ribozymes also catalyze reactions, with an important example being some parts of the ribosome. Synthetic molecules called artificial enzymes also display enzyme-like catalysis.

Enzyme activity can be affected by other molecules. Inhibitors are molecules that decrease enzyme activity; activators are molecules that increase activity. Many drugs and poisons are enzyme inhibitors. Activity is also affected by temperature, pressure, chemical environment (e.g., pH), and the concentration of substrate. Some enzymes are used commercially, for example, in the synthesis of antibiotics. In addition, some household products use enzymes to speed up biochemical reactions (e.g., enzymes in biological washing powders break down protein or fat stains on clothes; enzymes in meat tenderizers break down proteins into smaller molecules, making the meat easier to chew).

Structures and Mechanisms

Figure: *Ribbon diagram showing human carbonic anhydrase II. The grey sphere is the zinc cofactor in the active site. Diagram drawn from PDB 1MOO.*

Enzymes are in general globular proteins and range from just 62 amino acid residues in size, for the monomer of 4-oxalocrotonate tautomerase, to over 2,500 residues in the animal fatty acid synthase. A small number of RNA-based biological catalysts exist, with the most common being the ribosome; these are referred to as either RNA-enzymes or ribozymes.

The activities of enzymes are determined by their three-dimensional structure. However, although structure does determine function, predicting a novel enzyme's activity just from its structure is a very difficult problem that has not yet been solved.

Most enzymes are much larger than the substrates they act on, and only a small portion of the enzyme (around 2–4 amino acids) is directly involved in catalysis. The region that contains these catalytic residues, binds the substrate, and then carries out the reaction is known as the active site. Enzymes can also contain sites that bind cofactors, which are needed for catalysis. Some enzymes also have binding sites for small molecules, which are often direct or indirect products or substrates of the reaction catalyzed. This binding can serve to increase or decrease the enzyme's activity, providing a means for feedback regulation.

Like all proteins, enzymes are long, linear chains of amino acids that fold to produce a three-dimensional product. Each unique amino acid sequence produces a specific structure, which has unique properties. Individual protein chains may sometimes group together to form a protein complex. Most enzymes can be denatured—that is, unfolded and inactivated—by heating or chemical denaturants, which disrupt the three-dimensional structure of the protein. Depending on the enzyme, denaturation may be reversible or irreversible.

Structures of enzymes with substrates or substrate analogs during a reaction may be obtained using Time resolved crystallography methods.

Specificity

Enzymes are usually very specific as to which reactions they catalyze and the substrates that are involved in these reactions. Complementary shape, charge and hydrophilic/hydrophobic characteristics of enzymes and substrates are responsible for this specificity. Enzymes can also show impressive levels of stereospecificity, regioselectivity and chemoselectivity.

Some of the enzymes showing the highest specificity and accuracy are involved in the copying and expression of the genome. These enzymes have "proof-reading" mechanisms. Here, an enzyme such as DNA polymerase catalyzes a reaction in a first step and then checks that the product is correct in a second step. This two-step process results in average error rates of less than 1 error in 100 million reactions in high-fidelity mammalian polymerases. Similar proofreading mechanisms are also found in RNA polymerase, aminoacyl tRNA synthetases and ribosomes.

Whereas some enzymes have broad-specificity, as they can act on a relatively broad range of different physiologically relevant substrates, many enzymes possess small side activities which arose fortuitously (i.e. neutrally), which may be the starting point for the evolutionary selection of a new function; this phenomenon is known as enzyme promiscuity.

"Lock and key" Model

Enzymes are very specific, and it was suggested by Emil Fischer in 1894 that this was because both the enzyme and the substrate possess specific complementary geometric shapes that fit exactly into one another. This is often referred to as "the lock and key" model. However, while this model explains enzyme specificity, it fails to explain the stabilization of the transition state that enzymes achieve.

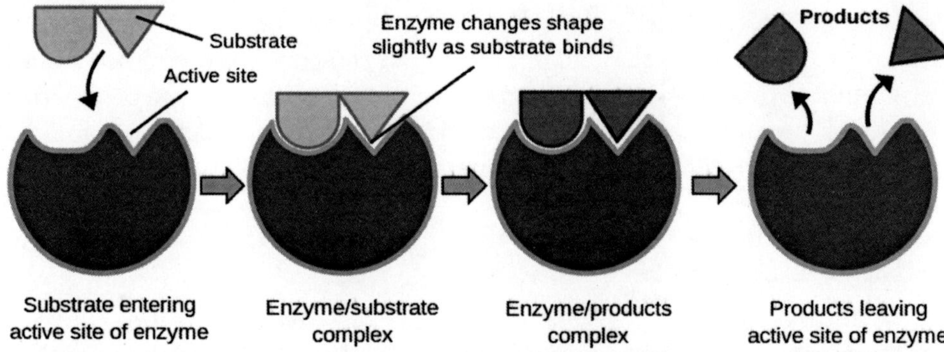

Figure: *Diagrams to show the induced fit hypothesis of enzyme action*

In 1958, Daniel Koshland suggested a modification to the lock and key model: since enzymes are rather flexible structures, the active site is continuously reshaped by interactions with the substrate as the substrate interacts with the enzyme. As a result, the substrate does not simply bind to a rigid active site; the amino acid side-chains that make up the active site are molded into the precise positions that enable the enzyme to perform its catalytic function. In some cases, such as glycosidases, the substrate molecule also changes shape slightly as it enters the active site. The active site continues to change until the substrate is completely bound, at which point the final shape and charge is determined. Induced fit may enhance the fidelity of molecular recognition in the presence of competition and noise via the conformational proofreading mechanism.

Based on Fischer's lock-and-key model and Koshland's induced fit theory, the Chou's distorted key theory for peptide drugs was proposed to develop peptide drugs against HIV/AIDS and SARS.

Mechanisms

Enzymes can act in several ways, all of which lower $\Delta G^{\ddagger}$ (Gibbs energy):

- Lowering the activation energy by creating an environment in which the transition state is stabilized (e.g. straining the shape of a substrate—by binding the transition-state conformation of the substrate/product molecules, the enzyme distorts the bound substrate(s) into their transition state form, thereby reducing the amount of energy required to complete the transition).
- Lowering the energy of the transition state, but without distorting the substrate, by creating an environment with the opposite charge distribution to that of the transition state.

- Providing an alternative pathway. For example, temporarily reacting with the substrate to form an intermediate ES complex, which would be impossible in the absence of the enzyme.
- Reducing the reaction entropy change by bringing substrates together in the correct orientation to react. Considering $\Delta H^{\ddagger}$ alone overlooks this effect.
- Increases in temperatures speed up reactions. Thus, temperature increases help the enzyme function and develop the end product even faster. However, if heated too much, the enzyme's shape deteriorates and the enzyme becomes denatured. Some enzymes like thermolabile enzymes work best at low temperatures.
- The function of a protein is dependent on its structure so that if the structure is disrupted, so is its function.

It is interesting that this entropic effect involves destabilization of the ground state, and its contribution to catalysis is relatively small.

Transition State Stabilization

The understanding of the origin of the reduction of $\Delta G^{\ddagger}$ requires one to find out how the enzymes can stabilize its transition state more than the transition state of the uncatalyzed reaction. It seems that the most effective way for reaching large stabilization is the use of electrostatic effects, in particular, when having a relatively fixed polar environment that is oriented towards the charge distribution of the transition state. Such an environment does not exist in the uncatalyzed reaction in water.

Cofactors and Coenzymes

Cofactors: Some enzymes do not need any additional components to show full activity. However, others require non-protein molecules called cofactors to be bound for activity. Cofactors can be either inorganic (e.g., metal ions and iron-sulfur clusters) or organic compounds (e.g., flavin and heme). Organic cofactors can be either prosthetic groups, which are tightly bound to an enzyme, or coenzymes, which are released from the enzyme's active site during the reaction. Coenzymes include NADH, NADPH and adenosine triphosphate. These molecules transfer chemical groups between enzymes.

An example of an enzyme that contains a cofactor is carbonic anhydrase, and is shown in the ribbon diagram above with a zinc cofactor bound as part of its active site. These tightly bound molecules are usually found in the active site and are involved in catalysis. For example, flavin and heme cofactors are often involved in redox reactions.

Enzymes that require a cofactor but do not have one bound are called *apoenzymes* or *apoproteins*. An apoenzyme together with its cofactor(s) is called a *holoenzyme* (this is the active form). Most cofactors are not covalently attached to an enzyme, but are very tightly bound. However, organic prosthetic groups can be covalently bound (e.g., biotin in the enzyme pyruvate carboxylase). The term "holoenzyme" can also be applied to enzymes that contain multiple protein subunits, such as the DNA polymerases; here the holoenzyme is the complete complex containing all the subunits needed for activity.

Coenzymes

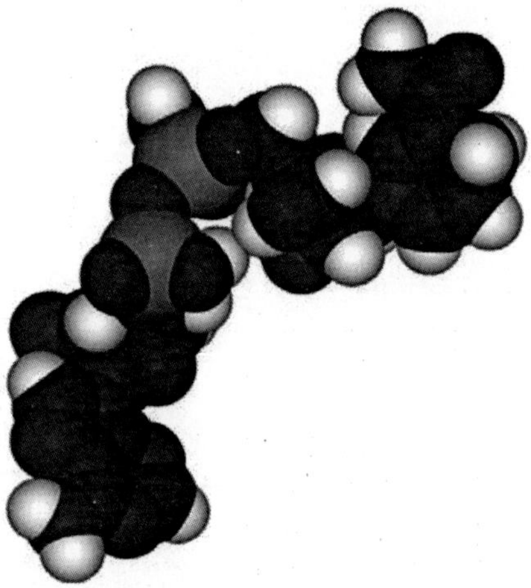

Figure: *Space-filling model of the coenzyme NADH*

Coenzymes are small organic molecules that can be loosely or tightly bound to an enzyme. Tightly bound coenzymes can be called prosthetic groups. Coenzymes transport chemical groups from one enzyme to another. Some of these chemicals such as riboflavin, thiamine and folic acid are vitamins (compounds that cannot be synthesized by the body and must be acquired from the diet). The chemical groups carried include the hydride ion (H^-) carried by NAD or $NADP^+$, the phosphate group carried by adenosine triphosphate, the acetyl group carried by coenzyme A, formyl, methenyl or methyl groups carried by folic acid and the methyl group carried by S-adenosylmethionine.

Since coenzymes are chemically changed as a consequence of enzyme action, it is useful to consider coenzymes to be a special class of substrates, or second substrates, which are common to many different enzymes. For example, about 700 enzymes are known to use the coenzyme NADH.

Coenzymes are usually continuously regenerated and their concentrations maintained at a steady level inside the cell: for example, NADPH is regenerated through the pentose phosphate pathway and

S-adenosylmethionine by methionine adenosyltransferase. This continuous regeneration means that even small amounts of coenzymes are used very intensively. For example, the human body turns over its own weight in ATP each day.

Biological Function

Enzymes serve a wide variety of functions inside living organisms. They are indispensable for signal transduction and cell regulation, often via kinases and phosphatases. They also generate movement, with myosin hydrolyzing ATP to generate muscle contraction and also moving cargo around the cell as part of the cytoskeleton. Other ATPases in the cell membrane are ion pumps involved in active transport. Enzymes are also involved in more exotic functions, such as luciferase generating light in fireflies. Viruses can also contain enzymes for infecting cells, such as the HIV integrase and reverse transcriptase, or for viral release from cells, like the influenza virus neuraminidase.

An important function of enzymes is in the digestive systems of animals. Enzymes such as amylases and proteases break down large molecules (starch or proteins, respectively) into smaller ones, so they can be absorbed by the intestines. Starch molecules, for example, are too large to be absorbed from the intestine, but enzymes hydrolyze the starch chains into smaller molecules such as maltose and eventually glucose, which can then be absorbed. Different enzymes digest different food substances. In ruminants, which have herbivorous diets, microorganisms in the gut produce another enzyme, cellulase, to break down the cellulose cell walls of plant fibre.

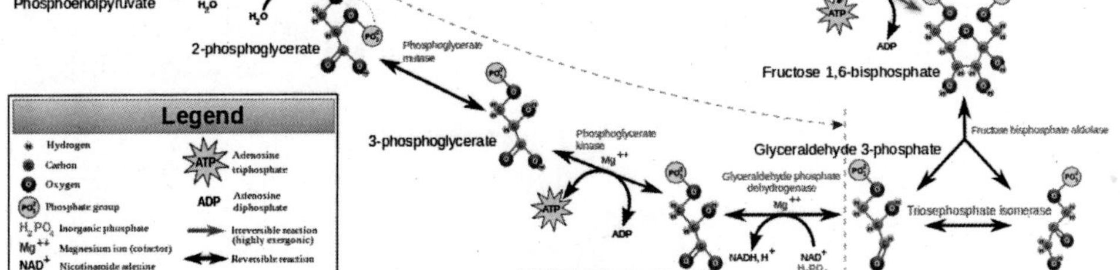

Figure: *Glycolytic enzymes and their functions in the metabolic pathway of glycolysis*

Several enzymes can work together in a specific order, creating metabolic pathways. In a metabolic pathway, one enzyme takes the product of another enzyme as a substrate. After the catalytic reaction, the product is then passed on to another enzyme. Sometimes more than one enzyme can catalyze the same reaction in parallel; this can allow more complex regulation: with, for example, a low constant activity provided by one enzyme but an inducible high activity from a second enzyme.

Enzymes determine what steps occur in these pathways. Without enzymes, metabolism would neither progress through the same steps nor be fast enough to serve the needs of the cell. Indeed, a metabolic pathway such as glycolysis could not exist independently of enzymes. Glucose, for example, can react directly with ATP to become phosphorylated at one or more of its carbons. In the absence of enzymes, this occurs so slowly as to be insignificant. However, if hexokinase is added, these slow reactions continue to take place except that phosphorylation at carbon 6 occurs so rapidly that, if the mixture is tested a short time later, glucose-6-phosphate is found to be the only significant product. As a consequence, the network of metabolic pathways within each cell depends on the set of functional enzymes that are present.

Enzymes in Soil

Soil enzymes increase the reaction rate at which plant residues decompose and release plant available nutrients. The substance acted upon by a soil enzyme is called the substrate. For example, glucosidase (soil enzyme) cleaves glucose from glucoside (substrate), a compound common in plants. Enzymes are specific to a substrate and have active sites that bind with the substrate to form a temporary complex. The enzymatic reaction releases a product, which can be a nutrient contained in the substrate.

Sources of soil enzymes include living and dead microbes, plant roots and residues, and soil animals. Enzymes stabilized in the soil matrix accumulate or form complexes with organic matter (humus), clay, and humus-clay complexes, but are no longer associated with viable cells. It is thought that 40 to 60% of enzyme activity can come from stabilized enzymes, so activity does not necessarily correlate highly with microbial biomass or respiration. Therefore, enzyme activity is the cumulative effect of long term microbial activity and activity of the viable population at sampling. However, an example of an enzyme that only reflects activity of viable cells is dehydrogenase, which in theory can only occur in viable cells and not in stabilized soil complexes.

Why it is important: Enzymes respond to soil management changes long before other soil quality indicator changes are detectable. Soil enzymes play an important role in organic matter decomposition and nutrient cycling. Some enzymes only facilitate the breakdown of organic matter (e.g., hydrolase, glucosidase), while others are involved in nutrient mineralization (e.g., amidase, urease, phosphatase, sulfates). With the exception of phosphatase activity, there is no strong evidence that directly relates enzyme activity to nutrient availability or crop production. The relationship may be indirect considering nutrient mineralization to plant available forms is accomplished with the contribution of enzyme activity.

***Table:** Role of soil enzymes*

Enzyme	***Organic Matter Substances Acted On***	***End Product***	***Significance***	***Predictor of Soil Function***
Beta glucosidase	carbon compounds	glucose (sugar)	energy for microorganisms	organic matter decomposition
FDA hydrolysis	organic matter	carbon and various nutrients	energy and nutrients for microorganisms, measure microbial biomass	organic matter decomposition nutrient cycling
Amidase	carbon and nitrogen compounds	ammonium (NH4)	plant available NH4	nutrient cycling
Urease	nitrogen (urea)	ammonia (NH3) and carbon dioxide (CO2)	plant available NH4	nutrient cycling
Phosphatase	phosphorus	phosphate (PO4)	plant available P	nutrient cycling
Sulfatase	sulfur	sulfate (SO4)	plant available S	nutrient cycling

Specific problems that might be caused by poor function: Absence or suppression of soil enzymes prevents or reduces processes that can affect plant nutrition. Poor enzyme activity (e.g., pesticide degrading enzymes) can result in an accumulation of chemicals that are harmful to the environment; some of these chemicals may further inhibit soil enzyme activity.

What you can do: Organic amendment applications, crop rotation, and cover crops have been shown to enhance enzyme activity (figures 1 and 2). The positive effect of pasture (figure 2) is associated with the input of animal manure and less soil disturbance. Agricultural methods that modify soil pH (e.g., liming) can also change enzyme activity.

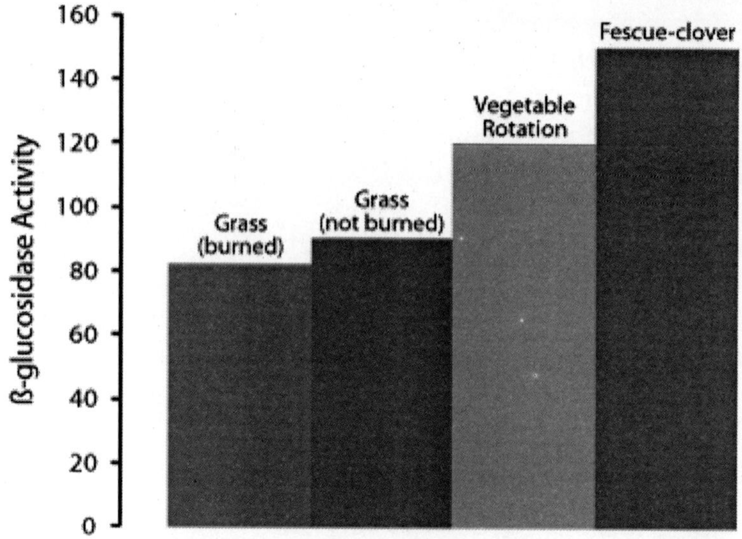

***Figure:** Effects of cropping systems on β-glucosidase (adapted from Dick 1994).*

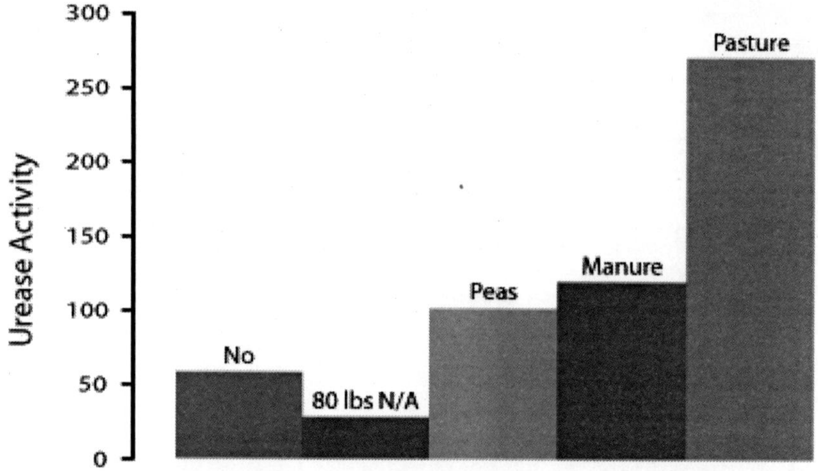

***Figure:** Effects of management on urease activity*

Measuring soil enzymes: Enzymes are measured indirectly by determining their activity in the laboratory using biochemical assays. Enzyme assays reflect potential activity and do not represent true in situ activity levels and must be viewed as an index.

When possible, compare the site of interest to samples taken from an adjacent, undisturbed site on the same soil type. Alternatively, for a newly implemented land management system, track changes from time zero to five or more years with annual sampling to detect temporal changes in activity of soil enzymes.

Importance of Soil to Agriculture

Figure: *We all know that soil is a key element of agriculture. Without it we wouldn´t be able to grow plants, which are used as food for both humans and animals.*

There are different definitions of soil, depending on the approach we want to take. For us (from the agricultural point of view) it is an unconsolidated mineral or organic material that is on the surface of the earth in which we grow plants. From a geological point of view, we would probably look at how it was formed and what kinds of layers it has. As far as food is concerned it is important to study a couple of its properties that are fundamental for plant growth. There are two important aspects as far as soil properties are concerned. First we focus on its hydraulic conductivity, the ability of the soil to manage, hold and drain water. Second, we look at its nutrient management, which in addition to the above-mentioned factors analyzes the organic matter content, cation exchange capacity and coatings on sand grains. All the above-mentioned properties determine the texture of soil. Every soil is a mixture

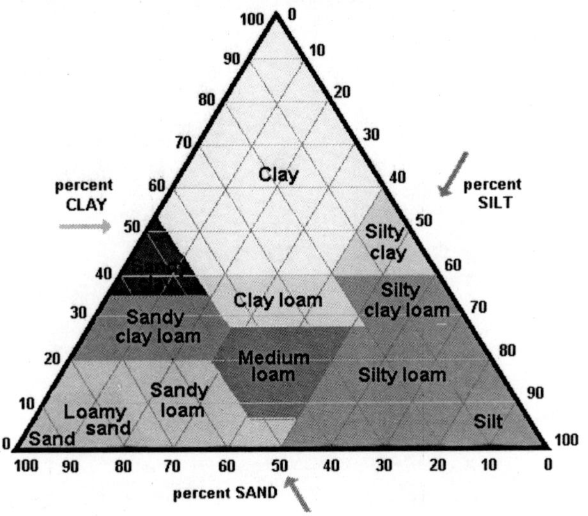

Components: sand, clay and silt. A very useful tool to verify a given soil texture is the soil textural triangle that you can see on the right. It shows you what kind of soil you are working on by taking into consideration the soil's components. Loam is the most fertile type of soil and has a mixture of these components.

For example, a more sandy soil will have a high hydraulic conductivity and a low water holding capacity. Water drains through sand very easily and if too much water enters this kind of soil it can lead to nitrate leaching which can be lethal to plants. Sandy soils need a special kind of irrigations systems. Most of farmers try to irrigate these soils many times throughout a day with small quantities of water. This helps the plants efficiently use the water supply without it washing away all the necessary nutrients.

There are different solutions to enhance sandy soil performance. One of them is to add more of the organic matter content. The organic matter is a mixture of living and recently dead materials such as previous crop residue, livestock waste or simple organic matter, such as leaves. The organic matter does two things for our soil. It adds water holding capacity and nutrient holding capacity. You can envision this process by comparing organic matter to a sponge. Of course, there is a limit to how much of the organic matter you can introduce to soil. If there is too much of it, in addition to supplying plants with too much nutrients it will start oxidizing, which provides an additional supply of nutrients. Soils that have a higher percentage of clay will have a lower hydraulic conductivity and a high water holding capacity. Sometimes when there is a compacted layer of clay underground it can lead to flooding after heavy storms. Water does not have anywhere to go in such a short time, so its level goes higher and eventually floods our plants. Moreover a higher percentage of clay and silt in the soil will increase the cation exchange capacity. It is an ability of the soil to hold positively charged nutrients called cations (calcium, magnesium, potassium and ammonium–N).

On our planet we have 12 different soil types known as orders. Each one of them needs a different treatment and farmers should know about these practices. How a farmer manages the farm can have crucial impacts on the environment. Many times, poor water management can lead to severe contamination of near by watersheds.

Agricultural Soil Science

Agricultural soil science is a branch of soil science that deals with the study of edaphic conditions as they relate to the production of food and fibre. In this context, it is also a constituent of the field of agronomy and is thus also described as soil agronomy. Prior to the development of pedology in the 19th century, agricultural soil science (or edaphology) was the only branch of soil science. The bias of early soil science towards viewing soils only in terms of their agricultural potential continues to define the soil science profession in both academic and popular settings as of 2006.

Current Status

Agricultural soil science follows the holistic method. Soil is investigated in relation to and as integral part of terrestrial ecosystems but is also recognized as a manageable natural resource. Agricultural soil science studies the chemical, physical, biological, and mineralogical composition of soils as they relate to agriculture. Agricultural soil scientists develop methods that will improve the use of soil and increase the production of food and fibre crops. Emphasis continues to grow on the importance of soil sustainability. Soil degradation such as erosion, compaction, lowered fertility, and contamination continue to be serious concerns. They conduct research in irrigation and drainage, tillage, soil classification, plant nutrition, soil fertility, and other areas.

Although maximizing plant (and thus animal) production is a valid goal, sometimes it may come at high cost which can be readily evident (e.g. massive crop disease stemming from monoculture) or long-term (e.g. impact of chemical fertilizers and pesticides on human health). An agricultural soil scientist may come up with a plan that can maximize production using sustainable methods and solutions, and in order to do that he must look into a number of science fields including agricultural science, physics, chemistry, biology, meteorology and geology.

Soil Variables

Some soil variables of special interest to agricultural soil science are:

- *Soil Texture or Soil Composition:* Soils are composed of solid particles of various sizes. In decreasing order, these particles are sand, silt and clay. Every soil can be classified according to the relative percentage of sand, silt and clay it contains.
- *Aeration and Porosity:* Atmospheric air contains elements such as oxygen, nitrogen, carbon and others. These elements are prerequisites for life on Earth. Particularly, all cells (including root cells) require oxygen to function and if conditions become anaerobic they fail to respire and metabolize. Aeration in this context refers to the mechanisms by which air is delivered to the soil. In natural ecosystems soil aeration is chiefly

accomplished through the vibrant activity of the biota. Humans commonly aerate the soil by tilling and plowing, yet such practice may cause degradation. Porosity refers to the air-holding capacity of the soil.

- *Drainage:* In soils of bad drainage the water delivered through rain or irrigation may pool and stagnate. As a result prevail anaerobic conditions and plant roots suffocate. Stagnant water also favours plant-attacking water molds. In soils of excess drainage, on the other hand, plants don't get to absorb adequate water and nutrients are washed from the porous medium to end up in groundwater reserves.
- *Water Content:* Without soil moisture there is no transpiration, no growth and plants wilt. Technically, plant cells loose their pressure. Plants contribute directly to soil moisture. For instance, they create a leafy cover that minimizes the evaporative effects of solar radiation. But even when plants or parts of plants die, the decaying plant matter produces a thick organic cover that protects the soil from evaporation, erosion and compaction.
- *Water Potential:* Water potential describes the tendency of the water to flow from one area of the soil to another. While water delivered to the soil surface normally flows downward due to gravity, at some point it meets increased pressure which causes a reverse upward flow. This effect is known as water suction.
- *Horizonation:* Typically found in advanced and mature soils, horizonation refers to the creation of soil layers with differing characteristics. It affects almost all soil variables.
- *Fertility:* A fertile soil is one rich in nutrients and organic matter. Modern agricultural methods have rendered much of the arable land infertile. In such cases, soil can no longer support on its own plants with high nutritional demand and thus needs an external source of nutrients. However there are cases where human activity is thought to be responsible for transforming rather normal soils into super-fertile ones.
- *Biota and Soil Biota:* Organisms interact with the soil and contribute to its quality in innumerable ways. Sometimes the nature of interaction may be unclear, yet a rule is becoming evident: The amount and diversity of the biota is "proportional" to the quality of the soil. Clades of interest include bacteria, fungi, nematodes, annelids and arthropods.
- *Soil Acidity or Soil pH and Cation-exchange Capacity:* Root cells act as hydrogen pumps and the surrounding concentration of

hydrogen ions affects their ability to absorb nutrients. pH is a measure of this concentration. Each plant species achieves maximum growth in a particular pH range, yet the vast majority of edible plants can grow in soil pH between 5.0 and 7.5.

Soil Fertility

Agricultural soil scientists study ways to make soils more productive. They classify soils and test them to determine whether they contain nutrients vital to plant growth. Such nutritional substances include compounds of nitrogen, phosphorus, and potassium. If a certain soil is deficient in these substances, fertilizers may provide them. Agricultural soil scientists investigate the movement of nutrients through the soil, and the amount of nutrients absorbed by a plant's roots. Agricultural soil scientists also examine the development of roots and their relation to the soil. Some agricultural soil scientists try to understand the structure and function of soils in relation to soil fertility. They grasp the structure of soil as porous solid. The solid frames of soil consist of mineral derived from the rocks and organic matter originated from the dead bodies of various organisms. The pore space of the soil is essential for the soil to become productive. Small pores serve as water reservoir supplying water to plants and other organisms in the soil during the rain-less period. The water in the small pores of soils is not pure water; they call it soil solution. In soil solution, various plant nutrients derived from minerals and organic matters in the soil are there. This is measured through the cation exchange capacity. Large pores serve as water drainage pipe to allow the excessive water pass through the soil, during the heavy rains. They also serve as air tank to supply oxygen to plant roots and other living beings in the soil. In short, agricultural soil scientists see the soil as a vessel, the most precious one for us, containing all of the substances needed by the plants and other living beings on earth.

Soil Preservation

In addition, agricultural soil scientists develop methods to preserve the agricultural productivity of soil and to decrease the effects on productivity of erosion by wind and water. For example, a technique called contour plowing may be used to prevent soil erosion and conserve rainfall. Researchers in agricultural soil science also seek ways to use the soil more effectively in addressing associated challenges. Such challenges include the beneficial reuse of human and animal wastes using agricultural crops; agricultural soil management aspects of preventing water pollution and the build-up in agricultural soil of chemical pesticides.

3

Lipids

Lipids are a group of naturally occurring molecules that include fats, waxes, sterols, fat-soluble vitamins (such as vitamins A, D, E, and K), monoglycerides, diglycerides, triglycerides, phospholipids, and others. The main biological functions of lipids include storing energy, signaling, and acting as structural components of cell membranes. Lipids have applications in the cosmetic and food industries as well as in nanotechnology.

Lipids may be broadly defined as hydrophobic or amphiphilic small molecules; the amphiphilic nature of some lipids allows them to form structures such as vesicles, liposomes, or membranes in an aqueous environment. Biological lipids originate entirely or in part from two distinct types of biochemical subunits or "building-blocks": ketoacyl and isoprene groups. Using this approach, lipids may be divided into eight categories: fatty acids, glycerolipids, glycerophospholipids, sphingolipids, saccharolipids, and polyketides (derived from condensation of ketoacyl subunits); and sterol lipids and prenol lipids (derived from condensation of isoprene subunits).

Although the term *lipid* is sometimes used as a synonym for fats, fats are a subgroup of lipids called triglycerides. Lipids also encompass molecules such as fatty acids and their derivatives (including tri-, di-, monoglycerides, and phospholipids), as well as other sterol-containing metabolites such as cholesterol. Although humans and other mammals use various biosynthetic pathways to both break down and synthesize lipids, some essential lipids cannot be made this way and must be obtained from the diet.

Categories of Lipids

Fatty Acids

Fatty acids, or fatty acid residues when they form part of a lipid, are a diverse group of molecules synthesized by chain-elongation of an acetyl-CoA primer with malonyl-CoA or methylmalonyl-CoA groups in a process called fatty acid synthesis. They are made of a hydrocarbon chain that terminates with a carboxylic acid group; this arrangement confers the molecule with a polar, hydrophilic end, and a nonpolar, hydrophobic end that is insoluble in water. The fatty acid structure is one of the most fundamental categories of biological lipids, and is commonly used as a building-block of more structurally complex lipids. The carbon chain, typically between four and 24 carbons long, may be saturated or unsaturated, and may be attached to functional groups containing oxygen, halogens, nitrogen, and sulfur. If a fatty acid contains a double bond, there is the possibility of either a *cis* or *trans* geometric isomerism, which significantly affects the molecule's configuration. *Cis*-double bonds cause the fatty acid chain to bend, an effect that is compounded with more double bonds in the chain. Three double bonds in 18-carbon *linolenic acid*, the most abundant fatty-acyl chains of plant *thylakoid membranes*, render these membranes highly *fluid* despite environmental low-temperatures, and also makes linolenic acid give dominating sharp peaks in high resolution 13-C NMR spectra of chlorplasts. This in turn plays an important role in the structure and function of cell membranes. Most naturally occurring fatty acids are of the *cis* configuration, although the *trans* form does exist in some natural and partially hydrogenated fats and oils.

Examples of biologically important fatty acids include the eicosanoids, derived primarily from arachidonic acid and eicosapentaenoic acid, that include prostaglandins, leukotrienes, and thromboxanes. Docosahexaenoic acid is also important in biological systems, particularly with respect to sight. Other major lipid classes in the fatty acid category are the fatty esters and fatty amides. Fatty esters include important biochemical intermediates such as wax esters, fatty acid thioester coenzyme A derivatives, fatty acid thioester ACP derivatives and fatty acid carnitines. The fatty amides include N-acyl ethanolamines, such as the cannabinoid neurotransmitter anandamide.

Glycerolipids

Glycerolipids are composed of mono-, di-, and tri-substituted glycerols, the most well-known being the fatty acid triesters of glycerol, called triglycerides. The word "triacylglycerol" is sometimes used synonymously with "triglyceride". In these compounds, the three hydroxyl groups of glycerol are each esterified, typically by different fatty acids. Because they function as an energy store, these lipids comprise the bulk of storage fat in animal tissues.

The hydrolysis of the ester bonds of triglycerides and the release of glycerol and fatty acids from adipose tissue are the initial steps in metabolising fat.

Additional subclasses of glycerolipids are represented by glycosylglycerols, which are characterized by the presence of one or more sugar residues attached to glycerol via a glycosidic linkage. Examples of structures in this category are the digalactosyldiacylglycerols found in plant membranes and seminolipid from mammalian sperm cells.

Glycerophospholipids

Glycerophospholipids, usually referred to as phospholipids, are ubiquitous in nature and are key components of the lipid bilayer of cells, as well as being involved in metabolism and cell signaling. Neural tissue (including the brain) contains relatively high amounts of glycerophospholipids, and alterations in their composition has been implicated in various neurological disorders. Glycerophospholipids may be subdivided into distinct classes, based on the nature of the polar headgroup at the *sn*-3 position of the glycerol backbone in eukaryotes and eubacteria, or the *sn*-1 position in the case of archaebacteria.

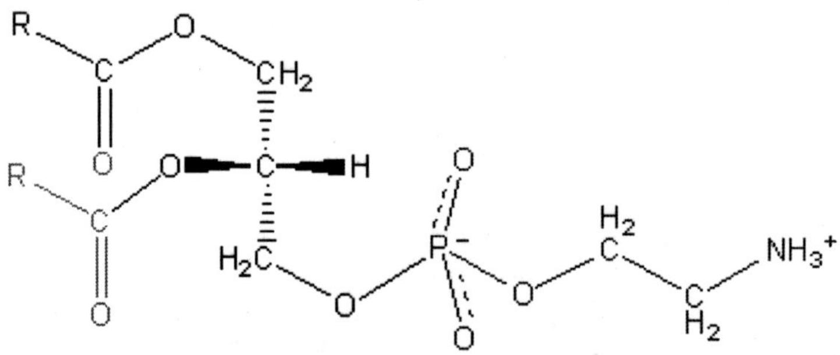

***Figure:** Phosphatidylethanolamine*

Examples of glycerophospholipids found in biological membranes are phosphatidylcholine (also known as PC, GPCho or lecithin), phosphatidylethanolamine (PE or GPEtn) and phosphatidylserine (PS or GPSer). In addition to serving as a primary component of cellular membranes and binding sites for intra- and intercellular proteins, some glycerophospholipids in eukaryotic cells, such as phosphatidylinositols and phosphatidic acids are either precursors of or, themselves, membrane-derived second messengers. Typically, one or both of these hydroxyl groups are acylated with long-chain fatty acids, but there are also alkyl-linked and 1Z-alkenyl-linked (plasmalogen) glycerophospholipids, as well as dialkylether variants in archaebacteria.

Sphingolipids

Sphingolipids are a complicated family of compounds that share a common structural feature, a sphingoid base backbone that is synthesized *de novo* from the amino acid serine and a long-chain fatty acyl CoA, then converted into ceramides, phosphosphingolipids, glycosphingolipids and other compounds. The major sphingoid base of mammals is commonly referred to as sphingosine. Ceramides (N-acyl-sphingoid bases) are a major subclass of sphingoid base derivatives with an amide-linked fatty acid. The fatty acids are typically saturated or mono-unsaturated with chain lengths from 16 to 26 carbon atoms.

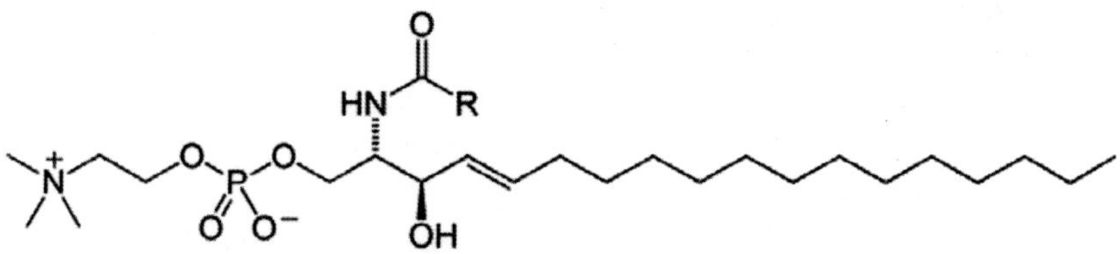

Figure: *Sphingomyelin*

The major phosphosphingolipids of mammals are sphingomyelins (ceramide phosphocholines), whereas insects contain mainly ceramide phosphoethanolamines and fungi have phytoceramide phosphoinositols and mannose-containing headgroups. The glycosphingolipids are a diverse family of molecules composed of one or more sugar residues linked via a glycosidic bond to the sphingoid base. Examples of these are the simple and complex glycosphingolipids such as cerebrosides and gangliosides.

Sterol Lipids

Sterol lipids, such as cholesterol and its derivatives, are an important component of membrane lipids, along with the glycerophospholipids and sphingomyelins. The steroids, all derived from the same fused four-ring core structure, have different biological roles as hormones and signaling molecules. The eighteen-carbon (C18) steroids include the estrogen family whereas the C19 steroids comprise the androgens such as testosterone and androsterone. The C21 subclass includes the progestogens as well as the glucocorticoids and mineralocorticoids. The secosteroids, comprising various forms of vitamin D, are characterized by cleavage of the B ring of the core structure. Other examples of sterols are the bile acids and their conjugates, which in mammals are oxidized derivatives of cholesterol and are synthesized in the liver. The plant equivalents are the phytosterols, such as β-sitosterol, stigmasterol, and brassicasterol; the latter compound is also used as a biomarker for algal growth. The predominant sterol in fungal cell membranes is ergosterol.

Prenol Lipids

Prenol lipids are synthesized from the five-carbon-unit precursors isopentenyl diphosphate and dimethylallyl diphosphate that are produced mainly via the mevalonic acid (MVA) pathway. The simple isoprenoids (linear alcohols, diphosphates, etc.) are formed by the successive addition of C5 units, and are classified according to number of these terpene units. Structures containing greater than 40 carbons are known as polyterpenes. Carotenoids are important simple isoprenoids that function as antioxidants and as precursors of vitamin A. Another biologically important class of molecules is exemplified by the quinones and hydroquinones, which contain an isoprenoid tail attached to a quinonoid core of non-isoprenoid origin. Vitamin E and vitamin K, as well as the ubiquinones, are examples of this class. Prokaryotes synthesize polyprenols (called bactoprenols) in which the terminal isoprenoid unit attached to oxygen remains unsaturated, whereas in animal polyprenols (dolichols) the terminal isoprenoid is reduced.

Saccharolipids

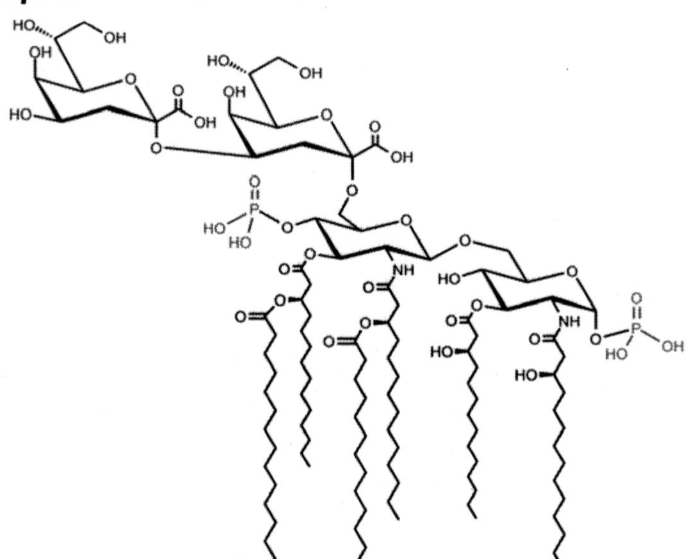

Figure: *Structure of the saccharolipid Kdo$_2$-Lipid A. Glucosamine residues in blue, Kdo residues in red, acyl chains in black and phosphate groups in green.*

Saccharolipids describe compounds in which fatty acids are linked directly to a sugar backbone, forming structures that are compatible with membrane bilayers. In the saccharolipids, a monosaccharide substitutes for the glycerol backbone present in glycerolipids and glycerophospholipids. The most familiar saccharolipids are the acylated glucosamine precursors of the Lipid A component of the lipopolysaccharides in Gram-negative bacteria. Typical lipid A molecules are disaccharides of glucosamine, which are derivatized with as many as seven fatty-acyl chains. The minimal lipopolysaccharide required for growth in *E. coli* is Kdo_2-Lipid A, a hexa-acylated disaccharide of glucosamine that is glycosylated with two 3-deoxy-D-manno-octulosonic acid (Kdo) residues.

Polyketides

Polyketides are synthesized by polymerization of acetyl and propionyl subunits by classic enzymes as well as iterative and multimodular enzymes that share mechanistic features with the fatty acid synthases. They comprise a large number of secondary metabolites and natural products from animal, plant, bacterial, fungal and marine sources, and have great structural diversity. Many polyketides are cyclic molecules whose backbones are often further modified by glycosylation, methylation, hydroxylation, oxidation, and/or other processes. Many commonly used anti-microbial, anti-parasitic, and anti-cancer agents are polyketides or polyketide derivatives, such as erythromycins, tetracyclines, avermectins, and antitumor epothilones.

Biological Functions

Membranes: Eukaryotic cells are compartmentalized into membrane-bound organelles that carry out different biological functions. The glycerophospholipids are the main structural component of biological membranes, such as the cellular plasma membrane and the intracellular membranes of organelles; in animal cells the plasma membrane physically separates the intracellular components from the extracellular environment. The glycerophospholipids are amphipathic molecules (containing both hydrophobic and hydrophilic regions) that contain a glycerol core linked to two fatty acid-derived “tails” by ester linkages and to one “head” group by a phosphate ester linkage. While glycerophospholipids are the major component of biological membranes, other non-glyceride lipid components such as sphingomyelin and sterols (mainly cholesterol in animal cell membranes) are also found in biological membranes. In plants and algae, the galactosyldiacylglycerols, and sulfoquinovosyldiacylglycerol, which lack a phosphate group, are important components of membranes of chloroplasts and related organelles and are the most abundant lipids in photosynthetic tissues, including those of higher plants, algae and certain bacteria.

Plant thylakoid membranes have the largest lipid component of a non-bilayer forming monogalactosyl diglyceride (MGDG), and little phospholipids; despite this unique lipid composition, chloroplast thylakoid membranes have been shown to contain a dynamic lipid-bilayer matrix as revealed by magnetic resonance and electron microscope studies. Bilayers have been found to exhibit high levels of birefringence, which can be used to probe the degree of order (or disruption) within the bilayer using techniques such as dual polarization interferometry and Circular dichroism.

A biological membrane is a form of lipid bilayer. The formation of lipid bilayers is an energetically preferred process when the glycerophospholipids described above are in an aqueous environment. This is known as the hydrophobic effect. In an aqueous system, the polar heads of lipids align towards the polar, aqueous environment, while the hydrophobic tails minimize their contact with water and tend to cluster together, forming a vesicle; depending on the concentration of the lipid, this biophysical interaction may result in the formation of micelles, liposomes, or lipid bilayers. Other aggregations are also observed and form part of the polymorphism of amphiphile (lipid) behaviour. Phase behaviour is an area of study within biophysics and is the subject of current academic research. Micelles and bilayers form in the polar medium by a process known as the hydrophobic effect. When dissolving a lipophilic or amphiphilic substance in a polar environment, the polar molecules (i.e., water in an aqueous solution) become more ordered around the dissolved lipophilic substance, since the polar molecules cannot form hydrogen bonds to the lipophilic areas of the amphiphile. So in an aqueous environment, the water molecules form an ordered "clathrate" cage around the dissolved lipophilic molecule.

Energy Storage

Triglycerides, stored in adipose tissue, are a major form of energy storage both in animals and plants. The adipocyte, or fat cell, is designed for continuous synthesis and breakdown of triglycerides in animals, with breakdown controlled mainly by the activation of hormone-sensitive enzyme lipase. The complete oxidation of fatty acids provides high caloric content, about 9 kcal/g, compared with 4 kcal/g for the breakdown of carbohydrates and proteins. Migratory birds that must fly long distances without eating use stored energy of triglycerides to fuel their flights.

Signaling

In recent years, evidence has emerged showing that lipid signaling is a vital part of the cell signaling. Lipid signaling may occur via activation of G protein-coupled or nuclear receptors, and members of several different lipid categories have been identified as signaling molecules and cellular messengers. These include sphingosine-1-phosphate, a sphingolipid derived from ceramide that is a potent messenger molecule involved in regulating calcium mobilization, cell growth, and apoptosis; diacylglycerol (DAG) and the phosphatidylinositol phosphates (PIPs), involved in calcium-mediated activation of protein kinase C; the prostaglandins, which are one type of fatty-acid derived eicosanoid involved in inflammation and immunity; the steroid hormones such as estrogen, testosterone and cortisol, which modulate a host of functions such as reproduction, metabolism and blood pressure; and the oxysterols such as 25-hydroxy-cholesterol that are liver X receptor agonists. Phosphatidylserine lipids are known to be involved in signaling for the phagocytosis of apoptotic cells and/or pieces of cells. They accomplish this by being exposed to the extracellular face of the cell membrane after the inactivation of flippases which place them exclusively on the cytosolic side and the activation of scramblases, which scramble the orientation of the phospholipids. After this occurs, other cells recognize the phosphatidylserines and phagocytosize the cells or cell fragments exposing them.

Other Functions

The "fat-soluble" vitamins (A, D, E and K) – which are isoprene-based lipids – are essential nutrients stored in the liver and fatty tissues, with a diverse range of functions. Acyl-carnitines are involved in the transport and metabolism of fatty acids in and out of mitochondria, where they undergo beta oxidation. Polyprenols and their phosphorylated derivatives also play important transport roles, in this case the transport of oligosaccharides across membranes. Polyprenol phosphate sugars and polyprenol diphosphate sugars function in extra-cytoplasmic glycosylation reactions, in extracellular polysaccharide biosynthesis (for instance, peptidoglycan polymerization in bacteria), and in eukaryotic protein N-glycosylation. Cardiolipins are a subclass of glycerophospholipids containing four acyl chains and three glycerol groups that are particularly abundant in the inner mitochondrial membrane. They are believed to activate enzymes involved with oxidative phosphorylation. Lipids also form the basis of steroid hormones.

Metabolism

The major dietary lipids for humans and other animals are animal and plant triglycerides, sterols, and membrane phospholipids. The process of lipid metabolism synthesizes and degrades the lipid stores and produces the structural and functional lipids characteristic of individual tissues.

Biosynthesis

In animals, when there is an oversupply of dietary carbohydrate, the excess carbohydrate is converted to triglycerides. This involves the synthesis of fatty acids from acetyl-CoA and the esterification of fatty acids in the production of triglycerides, a process called lipogenesis. Fatty acids are made by fatty acid synthases that polymerize and then reduce acetyl-CoA units. The acyl chains in the fatty acids are extended by a cycle of reactions that add the acetyl group, reduce it to an alcohol, dehydrate it to an alkene group and then reduce it again to an alkane group. The enzymes of fatty acid biosynthesis are divided into two groups, in animals and fungi all these fatty acid synthase reactions are carried out by a single multifunctional protein, while in plant plastids and bacteria separate enzymes perform each step in the pathway. The fatty acids may be subsequently converted to triglycerides that are packaged in lipoproteins and secreted from the liver.

The synthesis of unsaturated fatty acids involves a desaturation reaction, whereby a double bond is introduced into the fatty acyl chain. For example, in humans, the desaturation of stearic acid by stearoyl-CoA desaturase-1 produces oleic acid. The doubly unsaturated fatty acid linoleic acid as well as the triply unsaturated α-linolenic acid cannot be synthesized in mammalian tissues, and are therefore essential fatty acids and must be obtained from the diet.

Triglyceride synthesis takes place in the endoplasmic reticulum by metabolic pathways in which acyl groups in fatty acyl-CoAs are transferred to the hydroxyl groups of glycerol-3-phosphate and diacylglycerol.

Terpenes and isoprenoids, including the carotenoids, are made by the assembly and modification of isoprene units donated from the reactive precursors isopentenyl pyrophosphate and dimethylallyl pyrophosphate. These precursors can be made in different ways. In animals and archaea, the mevalonate pathway produces these compounds from acetyl-CoA, while in plants and bacteria the non-mevalonate pathway uses pyruvate and glyceraldehyde 3-phosphate as substrates. One important reaction that uses these activated isoprene donors is steroid biosynthesis. Here, the isoprene units are joined together to make squalene and then folded up and formed into a set of rings to make lanosterol. Lanosterol can then be converted into other steroids such as cholesterol and ergosterol.

Degradation

Beta oxidation is the metabolic process by which fatty acids are broken down in the mitochondria and/or in peroxisomes to generate acetyl-CoA. For the most part, fatty acids are oxidized by a mechanism that is similar to, but not identical with, a reversal of the process of fatty acid synthesis. That is, two-carbon fragments are removed sequentially from the carboxyl end of the acid after steps of dehydrogenation, hydration, and oxidation to form a beta-keto acid, which is split by thiolysis. The acetyl-CoA is then ultimately converted into ATP, CO_2, and H_2O using the citric acid cycle and the electron transport chain.

Hence the Krebs Cycle can start at acetyl-CoA when fat is being broken down for energy if there is little or no glucose available.

The energy yield of the complete oxidation of the fatty acid palmitate is 106 ATP. Unsaturated and odd-chain fatty acids require additional enzymatic steps for degradation.

Nutrition and Health

Most of the fat found in food is in the form of triglycerides, cholesterol, and phospholipids. Some dietary fat is necessary to facilitate absorption of fat-soluble vitamins (A, D, E, and K) and carotenoids. Humans and other mammals have a dietary requirement for certain essential fatty acids, such as linoleic acid (an omega-6 fatty acid) and alpha-linolenic acid (an omega-3 fatty acid) because they cannot be synthesized from simple precursors in the diet. Both of these fatty acids are 18-carbon polyunsaturated fatty acids differing in the number and position of the double bonds. Most vegetable oils are rich in linoleic acid (safflower, sunflower, and corn oils). Alpha-linolenic acid is found in the green leaves of plants, and in selected seeds, nuts, and legumes (in particular flax, rapeseed, walnut, and soy). Fish oils are particularly rich in the longer-chain omega-3 fatty acids eicosapentaenoic acid (EPA) and docosahexaenoic acid (DHA). A large number of studies have shown positive health benefits associated with consumption of omega-3 fatty acids on infant development, cancer, cardiovascular diseases, and various mental illnesses, such as depression, attention-deficit hyperactivity disorder, and dementia. In contrast, it is now well-established that consumption of trans fats, such as those present in partially hydrogenated vegetable oils, are a risk factor for cardiovascular disease.

A few studies have suggested that total dietary fat intake is linked to an increased risk of obesity and diabetes. However, a number of very large studies, including the Women's Health Initiative Dietary Modification Trial, an eight-year study of 49,000 women, the Nurses' Health Study and the Health Professionals Follow-up Study, revealed no such links. None of these studies suggested any connection between percentage of calories from fat and risk of cancer, heart disease, or weight gain. The Nutrition Source, a website maintained by the Department of Nutrition at the Harvard School of Public Health, summarizes the current evidence on the impact of dietary fat: "Detailed research—much of it done at Harvard—shows that the total amount of fat in the diet isn't really linked with weight or disease."

Lipid-phosphate Phosphatase

In enzymology, a lipid-phosphate phosphatase (EC 3.1.3.76) is an enzyme that catalyzes the chemical reaction (9S,10S)-10-hydroxy-9-(phosphonooxy)octadecanoate + H_2O (9S,10S)-9,10-dihydroxyoctadecanoate + phosphate

Thus, the two substrates of this enzyme are (9S,10S)-10-hydroxy-9-(phosphonooxy)octadecanoate and H_2O, whereas its two products are (9S,10S)-9,10-dihydroxyoctadecanoate and phosphate.

This enzyme belongs to the family of hydrolases, specifically those acting on phosphoric monoester bonds. The systematic name of this enzyme class is (9S,10S)-10-hydroxy-9-(phosphonooxy)octadecanoate phosphohydrolase. Other names in common use include hydroxy fatty acid phosphatase, dihydroxy fatty acid phosphatase, hydroxy lipid phosphatase, sEH (ambiguous), and soluble epoxide hydrolase (ambiguous).

Soil Biomass

Biomass is biological material derived from living, or recently living organisms. It most often refers to plants or plant-based materials which are specifically called lignocellulosic biomass. As an energy source, biomass can either be used directly via combustion to produce heat, or indirectly after converting it to various forms of biofuel. Conversion of biomass to biofuel can be achieved by different methods which are broadly classified into: *thermal*, *chemical*, and *biochemical* methods.

Wood remains the largest biomass energy source to date; examples include forest residues (such as dead trees, branches and tree stumps), yard clippings, wood chips and even municipal solid waste. In the second sense, biomass includes plant or animal matter that can be converted into fibres or other industrial chemicals, including biofuels. Industrial biomass can be grown from numerous types of plants, including miscanthus, switchgrass, hemp, corn, poplar, willow, sorghum, sugarcane, bamboo, and a variety of tree species, ranging from eucalyptus to oil palm (palm oil).

Plant energy is produced by crops specifically grown for use as fuel that offer high biomass output per hectare with low input energy. Some examples of these plants are wheat, which typically yield 7.5–8 tonnes of grain per hectare, and straw, which typically yield 3.5–5 tonnes per hectare in the UK. The grain can be used for liquid transportation fuels while the straw can be burned to produce heat or electricity. Plant biomass can also be degraded from cellulose to glucose through a series of chemical treatments, and the resulting sugar can then be used as a first generation biofuel.

Biomass can be converted to other usable forms of energy like methane gas or transportation fuels like ethanol and biodiesel. Rotting garbage, and agricultural and human waste, all release methane gas—also called "landfill gas" or "biogas." Crops, such as corn and sugar cane, can be fermented to produce the transportation fuel, ethanol. Biodiesel, another transportation fuel, can be produced from left-over food products like vegetable oils and animal fats. Also, biomass to liquids (BTLs) and cellulosic ethanol are still under research.

There is a great deal of research involving algal, or algae-derived, biomass due to the fact that it's a non-food resource and can be produced at rates 5 to 10 times faster than other types of land-based agriculture, such as corn and soy. Once harvested, it can be fermented to produce biofuels such as ethanol, butanol, and methane, as well as biodiesel and hydrogen.

The biomass used for electricity generation varies by region. Forest by-products, such as wood residues, are common in the United States. Agricultural waste is common in Mauritius (sugar cane residue) and Southeast Asia (rice husks). Animal husbandry residues, such as poultry litter, are common in the UK.

Biomass Sources

***Figure:** Eucalyptus in Brazil. Remains of the tree are reused for power generation.*

Historically, humans have harnessed biomass-derived energy since the time when people began burning wood to make fire. Even in today's modern era, biomass is the only source of fuel for domestic use in many developing countries. Biomass is all biologically-produced matter based in carbon, hydrogen and oxygen. The estimated biomass production in the world is 104.9 petagram ($104.9 * 10^{15}$ g) of carbon per year, about half in the ocean and half on land.

Wood remains the largest biomass energy source today; examples include forest residues (such as dead trees, branches and tree stumps), yard clippings, wood chips and even municipal solid waste. Wood energy is derived by using lignocellulosic biomass (second generation biofuels) as fuel. This is either using harvested wood directly as a fuel, or collecting from wood waste streams. The largest source of energy from wood is pulping liquor or "black liquor," a waste product from processes of the pulp, paper and paperboard industry. In the second sense, biomass includes plant or animal matter that can be converted into fibres or other industrial chemicals, including biofuels. Industrial biomass can be grown from numerous types of plants, including miscanthus, switchgrass, hemp, corn, poplar, willow, sorghum, sugarcane, bamboo, and a variety of tree species, ranging from eucalyptus to oil palm (palm oil).

Based on the source of biomass, biofuels are classified broadly into two major categories. First generation biofuels are derived from sources such as sugarcane and corn starch etc. Sugars present in this biomass are fermented to produce bioethanol, an alcohol fuel which furthermore can be used directly in a fuel cell to produce electricity or serve as an additive to gasoline. However, utilizing food based resource for fuel production aggravates food shortage problem. Second generation biofuels on the other hand utilize non-food based biomass sources such as agriculture and municipal waste. It mostly consists of lignocellulosic biomass which is not edible and is a low value waste for many industries. Despite being the favoured alternative, economical production of second generation biofuel is not yet achieved due to technological issues. These issues arise mainly due to chemical inertness and structural rigidity of lignocellulosic biomass.

Plant energy is produced by crops specifically grown for use as fuel that offer high biomass output per hectare with low input energy. Some examples of these plants are wheat, which typically yield 7.5–8 tons (tonnes?) of grain per hectare, and straw, which typically yield 3.5–5 tons (tonnes?) per hectare in the UK. The grain can be used for liquid transportation fuels while the straw can be burned to produce heat or electricity. Plant biomass can also be degraded from cellulose to glucose through a series of chemical treatments, and the resulting sugar can then be used as a first generation biofuel.

The main contributors of waste energy are municipal solid waste (MSW), manufacturing waste, and landfill gas. Energy derived from biomass is projected to be the largest non-hydroelectric renewable resource of electricity in the U.S between 2000 and 2020.

Biomass can be converted to other usable forms of energy like methane gas or transportation fuels like ethanol and biodiesel. Rotting garbage, and agricultural and human waste, all release methane gas—also called "landfill gas" or "biogas." Crops, such as corn and sugar cane, can be fermented to produce the transportation fuel, ethanol. Biodiesel, another transportation fuel, can be produced from left-over food products like vegetable oils and animal fats. Also, biomass to liquids (BTLs) and cellulosic ethanol are still under research.

There is a great deal of research involving algae, or algae-derived, biomass due to the fact that it's a non-food resource and can be produced at rates 5 to 10 times those of other types of land-based agriculture, such as corn and soy. Once harvested, it can be fermented to produce biofuels such as ethanol, butanol, and methane, as well as biodiesel and hydrogen. Efforts are being made to identify which species of algae are most suitable for energy production. Genetic engineering approaches could also be utilized to improve microalgae as a source of biofuel.

Soil Carbon

Soil carbon is the generic name for carbon held within the soil, primarily in association with its organic content. Soil carbon is the largest terrestrial pool of carbon, containing 2,200 gigatonnes (Gt) of it. Humans increasingly influence the size of this pool. Soil carbon plays a key role in the carbon cycle, and thus it is important in global climate models. Although the figure is frequently being revised upwards with new discoveries, over 2,700 gigatonnes (Gt) of carbon is stored in soils worldwide, which is well above the combined total of atmosphere (780 Gt) or biomass (575 Gt), most of which is wood. Carbon is taken out of the atmosphere by plant photosynthesis; about 60 Gt annually is incorporated into various types of soil organic matter (SOM), including surface litter; about 60 Gt annually is respired or oxidized from soil.

Soil carbon is the last major pool of the carbon cycle. The carbon that is fixed by plants is transferred to the soil via dead plant matter, including dead roots, leaves, and fruiting bodies. This dead organic matter creates a substrate which decomposes and respires back to the atmosphere as carbon dioxide or methane, depending on the availability of oxygen in the soil. Soil carbon is also oxidized by combustion and returned to the atmosphere as carbon dioxide. Soil carbon is primarily composed of biomass and non-biomass sources. Biomass carbon includes various bacteria and fungi. Non-biomass carbon sources or substrates reflect the chemical composition of plant biomass, and primarily include cellulose, starch, lignin, and other diverse organic carbon compounds. Some of the substrate carbon binds to the mineral soil, becoming encapsulated in soil aggregates (singular masses of coherent soil particles, or peds) or chemical complexional.

Biomass feeds on the substrate carbon compounds at different rates. Some of the carbon compounds are easily digested and respired by the microbes, resulting in a relatively short residence time. Others—like lignin, humic acid, or substrate encapsulated in soil aggregates—are very difficult for the biomass to absorb and have long residence times.

Soil Carbon and Soil Health

It is widely accepted that the carbon content of soil is a major factor in its overall health. Soil carbon improves the physical properties of soil. It increases the cation-exchange capacity (CEC) and water-holding capacity of sandy soil, and it contributes to the structural stability of clay soils by helping to bind particles into aggregates. SOM, of which carbon is a major part, holds a great proportion of nutrients cations and trace elements that are of importance to plant growth. It prevents nutrient leaching, and is integral to the organic acids that make minerals available to plants. It also buffers soil from strong changes in pH.

Losses of Soil Carbon

Although exact quantities are difficult to measure, human activities have caused massive losses of soil organic carbon. First was the use of fire, which removes soil cover and leads to immediate and continuing losses of soil organic carbon. Tillage and drainage both expose soil organic matter to oxygen and oxidation. In the Netherlands, East Anglia, Florida, and the California Delta, subsidence of peat lands from oxidation has been severe as a result of tillage and drainage. Grazing management that exposes soil (through either excessive or insufficient recovery periods) can also cause losses of soil organic carbon.

Managing Soil Carbon

Natural variations in SOM occur as a result of climate, organisms, parent material, time, and relief. The greatest contemporary influence has been that of humans; for example, historical SOM in Australian agricultural soils may have been twice the present range that is typically from 1.6 to 4.6 per cent. It has long been encouraged that farmers adjust practices to maintain or increase the organic component in the soil. On one hand, practices that hasten oxidation of carbon (such as burning crop stubbles or over-cultivation) are discouraged; on the other hand, incorporation of organic material (such as in manuring) has been encouraged. Increasing soil carbon is not a straightforward matter; it is made complex by the relative activity of soil biota, which can consume and release carbon and are made more active by the addition of nitrogen fertilizers.

Data Available on Soil Organic Carbon

Europe: The most homogeneous and comprehensive data on the organic carbon/matter content of European soils remain those that can be extracted and/or derived from the European Soil Database in combination with associated databases on land cover, climate, and topography. The modelled data refer to carbon content (%) in the surface horizon of soils in Europe. In an inventory on available national datasets, seven member states of the European Union have available datasets on organic carbon. In the article "Estimating soil organic carbon in Europe based on data collected through an European network" (Ecological Indicators 24, pp. 439-450), a comparison of national data with modelled data is performed. Finally, a new proposed model for estimation of soil organic carbon in agricultural soils has estimated current top SOC stock of 17.63 Gt in EU agricultural soils.

Managing for Catchment Health

Much of the contemporary literature on soil carbon relates to its role, or potential, as an atmospheric carbon sink to offset climate change. Despite this emphasis, a much wider range of soil and catchment health aspects are improved as soil carbon is increased. These benefits are difficult to quantify, due to the complexity of natural resource systems and the interpretation of what constitutes soil health; nonetheless, several benefits are proposed in the following points:

- Reduced erosion, sedimentation: increased soil aggregate stability means greater resistance to erosion; mass movement is less likely when soils are able to retain structural strength under greater moisture levels.
- Greater productivity: healthier and more productive soils can contribute to positive socio-economic circumstances.
- Cleaner waterways, nutrients and turbidity: nutrients and sediment tend to be retained by the soil rather than leach or wash off, and are so kept from waterways.

- Water balance: greater soil water holding capacity reduces overland flow and recharge to groundwater; the water saved and held by the soil remains available for use by plants.
- Climate change: Soils have the ability to retain carbon that may otherwise exist as atmospheric CO_2 and contribute to global warming.
- Greater biodiversity: soil organic matter contributes to the health of soil flora and, accordingly, the natural links with biodiversity in the greater biosphere.

Forest Soils

Forest soils constitute a large pool of carbon. Anthropogenic activities such as deforestation cause releases of carbon from this pool, which may significantly increase the concentration of greenhouse gas (GHG) in the atmosphere. Under the United Nations Framework Convention on Climate Change (UNFCCC), countries must estimate and report GHG emissions and removals, including changes in carbon stocks in all five pools (above- and below-ground biomass, dead wood, litter, and soil carbon) and associated emissions and removals from land use, land-use, change and forestry activities, according to the Intergovernmental Panel on Climate Change's good practice guidance. Tropical deforestation represents nearly 25 percent of total anthropogenic GHG emissions worldwide. Deforestation, forest degradation, and changes in land management practices can cause releases of carbon from soil to the atmosphere. For these reasons, reliable estimates of soil organic carbon stock and stock changes are needed for Reducing Emissions from Deforestation and Forest Degradation and GHG reporting under the UNFCCC.

The government of Tanzania—together with the Food and Agriculture Organization of the United Nations and the financial support of the government of Finland—have implemented a forest soil carbon monitoring programme to estimate soil carbon stock, using both survey and modelling-based methods.

Biochar

Biochar is a name for charcoal when it is used for particular purposes, especially as a soil amendment. Like all charcoal, biochar is created by pyrolysis of biomass. Biochar is under investigation as an approach to carbon sequestration to produce negative carbon dioxide emissions. Biochar thus has the potential to help mitigate climate change, via carbon sequestration. Independently, biochar can increase soil fertility, increase agricultural productivity, and provide protection against some foliar and soil-borne diseases. Furthermore, biochar reduces pressure on forests. Biochar is a stable solid, rich in carbon and can endure in soil for thousands of years.

Pre-Columbian Amazonians are believed to have used biochar to enhance soil productivity. They produced it by smoldering agricultural waste (i.e., covering burning biomass with soil) in pits or trenches. European settlers called it *terra preta de Indio*. Following observations and experiments, a research team working in French Guiana hypothesized that the Amazonian earthworm *Pontoscolex corethrurus* was the main agent of fine powdering and incorporation of charcoal debris to the mineral soil.

The term "biochar" was coined by Peter Read to describe charcoal used as a soil improvement.

Production

Pyrolysis produces biochar, liquids, and gases from biomass by heating the biomass in a low/no oxygen environment. The absence of oxygen prevents combustion. The relative yield of products from pyrolysis varies with temperature. Temperatures of 400–500 °C (752–932 °F) produce more char, while temperatures above 700 °C (1,292 °F) favour the yield of liquid and gas fuel components. Pyrolysis occurs more quickly at the higher temperatures, typically requiring seconds instead of hours. High temperature pyrolysis is also known as gasification, and produces primarily syngas. Typical yields are 60% bio-oil, 20% biochar, and 20% syngas. By comparison, slow pyrolysis can produce substantially more char (~50%). Once initialized, both processes produce net energy. For typical inputs, the energy required to run a "fast" pyrolyzer is approximately 15% of the energy that it outputs. Modern pyrolysis plants can use the syngas created by the pyrolysis process and output 3–9 times the amount of energy required to run.

The Amazonian pit/trench method harvests neither bio-oil nor syngas, and releases a large amount of CO_2, black carbon, and other greenhouse gases (GHG)s (and potentially, toxins) into the air. Commercial-scale systems process agricultural waste, paper byproducts, and even municipal waste and typically eliminate these side effects by capturing and using the liquid and gas products.

Centralized, Decentralized, and Mobile Systems

In a centralized system, all biomass in a region is brought to a central plant for processing. Alternatively, each farmer or group of farmers can operate a lower-tech kiln. Finally, a truck equipped with a pyrolyzer can move from place to place to pyrolyze biomass. Vehicle power comes from the syngas stream, while the biochar remains on the farm. The biofuel is sent to a refinery or storage site. Factors that influence the choice of system type include the cost of transportation of the liquid and solid byproducts, the amount of material to be processed, and the ability to feed directly into the power grid.

For crops that are not exclusively for biochar production, the residue-to-product ratio (RPR) and the collection factor (CF) the percent of the residue not used for other things, measure the approximate amount of feedstock that can be obtained for pyrolysis after harvesting the primary product. For instance, Brazil harvests approximately 460 million tons (MT) of sugarcane annually, with an RPR of 0.30, and a CF of 0.70 for the sugarcane tops, which normally are burned in the field. This translates into approximately 100 MT of residue annually which could be pyrolyzed to create energy and soil additives. Adding in the bagasse (sugarcane waste) (RPR=0.29 CF=1.0) which is otherwise burned (inefficiently) in boilers, raises the total to 230 MT of pyrolysis feedstock. Some plant residue, however, must remain on the soil to avoid increased costs and emissions from nitrogen fertilizers.

Pyrolysis technologies for processing loose and leafy biomass produce both biochar and syngas.

Thermo-catalytic Depolymerization

Alternatively, *thermo-catalytic depolymerization* using microwaves has recently been used to efficiently convert organic matter to biochar on an industrial scale, producing ~50% char.

Uses

Carbon Sink: The burning and natural decomposition of biomass and in particular agricultural waste adds large amounts of CO_2 to the atmosphere. Biochar that is stable, fixed, and 'recalcitrant' carbon can store large amounts of greenhouse gases in the ground for centuries, potentially reducing or stalling the growth in atmospheric greenhouse gas levels; at the same time its presence in the earth can improve water quality, increase soil fertility, raise agricultural productivity, and reduce pressure on old-growth forests.

Biochar can sequester carbon in the soil for hundreds to thousands of years, like coal. Such a carbon-negative technology would lead to a net withdrawal of CO_2 from the atmosphere, while producing and consuming energy." This technique is advocated by prominent scientists such as James Hansen, head of the NASA Goddard Institute for Space Studies, and James Lovelock, creator of the Gaia hypothesis, for mitigation of global warming by greenhouse gas remediation.

Researchers have estimated that sustainable use of biocharring could reduce the global net emissions of carbon dioxide (CO_2), methane, and nitrous oxide by up to 1.8 Pg CO_2-C equivalent (CO_2-C_e) per year (12% of current anthropogenic CO_2-Ce emissions; 1 Pg=1 Gt), and total net emissions over the course of the next century by 130 Pg CO_2-C_e, without endangering food security, habitat, or soil conservation.

Biochar is a high-carbon, fine-grained residue which today is produced through modern pyrolysis processes. Pyrolysis is the direct thermal decomposition of biomass in the absence of oxygen to obtain an array of solid (biochar), liquid (bio-oil), and gas (syngas) products. The specific yield from the pyrolysis is dependent on process conditions, and can be optimized to produce either energy or biochar.

Soil Amendment

For plants that require high potash and elevated pH, biochar can be used as a soil amendment to improve yield. Biochar can improve water quality, reduce soil emissions of greenhouse gases, reduce nutrient leaching, reduce soil acidity, and reduce irrigation and fertilizer requirements. Biochar was also found under certain circumstances to induce plant systemic responses to foliar fungal diseases and to improve plant responses to diseases caused by soilborne pathogens.

The various impacts of biochar can be dependent on the properties of the biochar, as well as the amount applied, and there is still a lack of knowledge about the important mechanisms and properties. Biochar impact may depend on regional conditions including soil type, soil condition (depleted or healthy), temperature, and humidity. Modest additions of biochar to soil reduce nitrous oxide N_2O emissions by up to 80% and eliminate methane emissions, which are both more potent greenhouse gases than CO_2.

Pollutants such as metals and pesticides seep into soil and contaminate food supplies, reducing the amount of land suitable for agricultural production. Studies have reported positive effects from biochar on crop production in degraded and nutrient–poor soils. Biochar can be designed with specific qualities to target distinct properties of soils. Biochar reduces leaching of critical nutrients, creates a higher crop uptake of nutrients, and provides greater soil availability of nutrients. At 10% levels biochar reduced contaminant levels in plants by up to 80%, while reducing total chlordane and DDX content in the plants by 68 and 79%, respectively. On the other hand, because of its high adsorption capacity, biochar may reduce the efficacy of soil applied pesticides that are needed for weed and pest control. High surface area biochars may be particularly problematic in this regard; more research into the long term effects of biochar addition to soil is needed.

Slash and Char

Switching from *slash and burn* to *slash and char* farming techniques in Brazil can decrease both deforestation of the Amazon basin and carbon dioxide emission, as well as increase crop yields. Slash and burn leaves only 3% of the carbon from the organic material in the soil.

Slash and char can keep up to 50% of the carbon in a highly stable form. Returning the biochar into the soil rather than removing it all for energy production reduces the need for nitrogen fertilizers, thereby reducing cost and emissions from fertilizer production and transport. Additionally, by improving the soil's ability to be tilled, fertility, and productivity, biochar–enhanced soils can indefinitely sustain agricultural production, whereas non-enriched soils quickly become depleted of nutrients, forcing farmers to abandon the fields, producing a continuous slash and burn cycle and the continued loss of tropical rainforest. Using pyrolysis to produce bio-energy also has the added benefit of not requiring infrastructure changes the way processing biomass for cellulosic ethanol does. Additionally, the biochar produced can be applied by the currently used machinery for tilling the soil or equipment used to apply fertilizer.

Water Retention

Biochar is a desirable soil material in many locations due to its ability to attract and retain water. This is possible because of its porous structure and high surface area. As a result, nutrients, phosphorus, and agrochemicals are retained for the plants benefit. Plants therefore, are healthier and fertilizers leach less into surface or groundwater.

Energy Production: Bio-oil and Syngas

Bio-oil can be used as a replacement for numerous applications where fuel oil is used, including fuelling space heaters, furnaces, and boilers. Additionally, these biofuels can be used to fuel some combustion turbines and reciprocating engines, and as a source to create several chemicals. If bio-oil is used without modification, care must be taken to prevent emissions of black carbon and other particulates. Syngas and bio-oil can also be "upgraded" to transportation fuels such as biodiesel and gasoline substitutes. If biochar is used for the production of energy rather than as a soil amendment, it can be directly substituted for any application that uses coal. Pyrolysis also may be the most cost-effective way of producing electrical energy from biomaterial. Syngas can be burned directly, used as a fuel for gas engines and gas turbines, converted to clean diesel fuel through the Fischer–Tropsch process, or potentially used in the production of methanol and hydrogen.

Bio-oil has a much higher energy density than the raw biomass material. Mobile pyrolysis units can be used to lower the costs of transportation of the biomass if the biochar is returned to the soil and the syngas stream is used to power the process. Bio-oil contains organic acids that are corrosive to steel containers, has a high water vapour content that is detrimental to ignition, and, unless carefully cleaned, contains some biochar particles which can block injectors. The greatest potential for bio-oil seems to be its use in a bio-refinery, where compounds that are valuable chemicals, pesticides, pharmaceuticals, or food additives are first extracted, and the remainder is either upgraded to fuel or reformed to syngas.

Direct and Indirect Benefits

- The pyrolysis of forest- or agriculture-derived biomass residue generates a biofuel without competition with crop production.
- Biochar is a pyrolysis byproduct that may be ploughed into soils in crop fields to enhance their fertility and stability, and for medium- to long-term carbon sequestration in these soils.
- Biochar enhances the natural process: the biosphere captures CO_2, especially through plant production, but only a small portion is stably sequestered for a relatively long time (soil, wood, etc.).
- Biomass production to obtain biofuels and biochar for carbon sequestration in the soil is a carbon-negative process, i.e. more CO_2 is removed from the atmosphere than released, thus enabling long-term sequestration.

Research

Intensive research into manifold aspects involving the pyrolysis/ biochar platform is underway around the world. From 2005-2012, there were 1,038 articles that included the word "biochar" or "bio-char" in the topic that had been indexed in the ISI Web of Science. Further research is in progress by such diverse institutions around the world as Cornell University, the University of Edinburgh, which has a dedicated research unit., and the Agricultural Research Organization (ARO) of Israel, Volcani Centre, where a network of researchers involved in biochar research (iBRN, Israel Biochar Researchers Network) was established as early as 2009.

Students at Stevens Institute of Technology in New Jersey are developing supercapacitors that use electrodes made of biochar. A process developed by University of Florida researchers that removes phosphate from water, also yields methane gas usable as fuel and phosphate-laden carbon suitable for enriching soil.

Emerging Commercial Sector

Calculations suggest that emissions reductions can be 12–84% greater if biochar is put back into the soil instead of being burned to offset fossil-fuel use. Thus Biochar sequestration offers the chance to turn bioenergy into a carbon-negative industry.

Johannes Lehmann, of Cornell University, estimates that pyrolysis can be cost-effective for a combination of sequestration and energy production when the cost of a CO_2 ton reaches \$37. As of mid-February 2010, CO_2 is trading at \$16.82/ton on the European Climate Exchange (ECX), so using pyrolysis for bioenergy production may be feasible even if it is more expensive than fossil fuel.

Current biochar projects make no significant impact on the overall global carbon budget, although expansion of this technique has been advocated as a geoengineering approach. In May 2009, the Biochar Fund received a grant from the Congo Basin Forest Fund for a project in Central Africa to simultaneously slow down deforestation, increase the food security of rural communities, provide renewable energy and sequester carbon.

Testing

The microbial biomass of soil is defined as the part of the organic matter in the soil that constitutes living microorganisms smaller than the 5-10 um^3. It is generally expressed in the milligrams of carbon per kilogram of soil or micrograms of carbon per gram of dry weight of soil. Typical biomass carbon ranges from 1 to 5% of soil organic matter. The degradation of organic compounds, such as industrial chemicals and pesticides, can be monitored by following changes in the soil microbial biomass.

Several methods can and are used to estimate the soil microbial biomass population. They can be summarized as follows:

1. Staining and counting the microbial cells.
2. The use of physiological parametres such as ATP, respiration, and heat output.
3. An application of a soil fumigate and measurement.

AGVISE currently employs a soil fumigate method for measuring the biomass of the soil. The soil is fumigated with chloroform to kill the microbial population. After the microbes are killed by fumigation, cytoplasm is released into the soil environment. The soil microbial biomass carbon is extracted with potassium sulfate on both fumigated and unfumigated soil. The carbon content of the extract is tested and the biomass is calculated based on the difference between the carbon content of fumigated vs. the unfumigated soil. We currently measure the carbon by dichromate oxidation, but plan on using our automatic carbon analyzer with an IR detector in the future.

A biomass reading is a snapshot picture of the microbial population on the day the sample was tested in the laboratory. It is not a stable number like the CEC or the texture of a soil. The biomass number will change with the environment in which the soil undergoes. After the soil is removed from a field setting, the biomass can and will change with storage conditions of the soil. Recommended storage conditions for biomass samples is 4°C. Freezing of soil samples is not recommended due to the adverse biocidal effects on the soil microbial biomass. If samples must be frozen, they should be preincubated for 7 to 10 days prior to testing. The drying of soil samples for biomass readings must be strictly avoided. Also, after the sampling procedure takes place, samples should not be exposed to direct sunlight.

Biomass of Samplesas Related to Texture

Soil Texture(USDA)	*% OM(mean)*	*MicrobialBiomassug/g*
Sand	2.0	55
Loamy Sand	1.5	137
Sandy Loam	1.6	106
Silt Loam	3.2	292
Loam	4.5	358

Biomass of Samplesas Related To OM

OrganicMatterRange	*AverageMicrobial Biomass ug/g*	*MicrobialBiomass Range ug/g*
0 to 1.0	76	10 to 165
1.0 to 2.0	130	17 to 379
2.0 to 3.0	169	24 to 418
3.0 to 4.0	219	119 to 300
4.0 to 5.0	345	127 to 454
5.0 to 6.0	427	369 to 506
6.0+	613	421 to 805

4

Carbohydrates, Nucleic Acid and Phenolic Compounds in Soils

Carbohydrates

Carbohydrates are the most abundant class of organic compounds found in living organisms. They originate as products of photosynthesis, an endothermic reductive condensation of carbon dioxide requiring light energy and the pigment chlorophyll.

$$n\ CO_2 + n\ H_2O + \text{energy} \rightarrow C_nH_{2n}O_n + n\ O_2$$

As noted here, the formulas of many carbohydrates can be written as carbon hydrates, $C_n(H_2O)_n$, hence their name. The carbohydrates are a major source of metabolic energy, both for plants and for animals that depend on plants for food. Aside from the sugars and starches that meet this vital nutritional role, carbohydrates also serve as a structural material (cellulose), a component of the energy transport compound ATP, recognition sites on cell surfaces, and one of three essential components of DNA and RNA.

Complexity	*Simple Carbohydrates*	*Complex*
	monosaccharides	Carbohydratesdisaccharides, oligosaccharides& polysaccharides

Size	TetroseC_4 sugars	PentoseC_5 sugars	HexoseC_6 sugars	HeptoseC_7 sugars	etc.

C=O Function	Aldose sugars having an aldehyde function or an acetal equivalent. Ketose sugars having a ketone function or an acetal equivalent.
Reactivity	Reducing sugars oxidized by Tollens' reagent (or Benedict's or Fehling's reagents).Non-reducing sugars not oxidized by Tollens' or other reagents.

Glucose

Carbohydrates have been given non-systematic names, although the suffix ose is generally used. The most common carbohydrate is glucose ($C_6H_{12}O_6$). Applying the terms defined above, glucose is a monosaccharide, an aldohexose (note that the function and size classifications are combined in one word) and a reducing sugar. The general structure of glucose and many other aldohexoses was established by simple chemical reactions. The following diagram illustrates the kind of evidence considered, although some of the reagents shown here are different from those used by the original scientists.

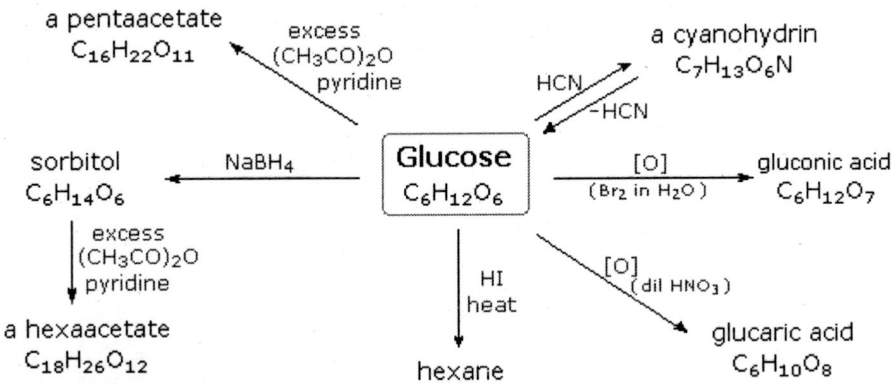

Hot hydriodic acid (HI) was often used to reductively remove oxygen functional groups from a molecule, and in the case of glucose this treatment gave hexane (in low yield). From this it was concluded that the six carbons are in an unbranched chain. The presence of an aldehyde carbonyl group was deduced from cyanohydrin formation, its reduction to the hexa-alcohol sorbitol, also called glucitol, and mild oxidation to the mono-carboxylic acid, glucuronic acid. Somewhat stronger oxidation by dilute nitric acid gave the diacid, glucaric acid, supporting the proposal of a six-carbon chain. The five oxygens remaining in glucose after the aldehyde was accounted for were thought to be in hydroxyl groups, since a penta-acetate derivative could be made. These hydroxyl groups were assigned, one each, to the last five carbon atoms, because geminal hydroxyl groups are normally unstable relative to the carbonyl compound formed by loss of water. Glucose and other saccharides are extensively cleaved by periodic acid, thanks to the abundance of vicinal diol moieties in their structure. This oxidative cleavage, known as the Malaprade reaction is particularly useful for the analysis of selective O-substituted derivatives of saccharides, since ether functions do not react. The stoichiometry of aldohexose cleavage is shown in the following equation.

$$HOCH_2(CHOH)_4CHO + 5\ HIO_4 \longrightarrow H_2C{=}O + 5\ HCO_2H + 5\ HIO_3$$

The Configuration of Glucose

The four chiral centres in glucose indicate there may be as many as sixteen (2^4) stereoisomers having this constitution. These would exist as eight diastereomeric pairs of enantiomers, and the initial challenge was to determine which of the eight corresponded to glucose. This challenge was accepted and met in 1891 by the German chemist Emil Fischer. His successful negotiation of the stereochemical maze presented by the aldohexoses was a logical tour de force, and it is fitting that he received the 1902 Nobel Prize for chemistry for this accomplishment. One of the first tasks faced by Fischer was to devise a method of representing the configuration of each chiral centre in an unambiguous manner. To this end, he invented a simple technique for drawing chains of chiral centres, that we now call the Fischer projection formula. Click on this link for a review.

At the time Fischer undertook the glucose project it was not possible to establish the absolute configuration of an enantiomer. Consequently, Fischer made an arbitrary choice for (+)-glucose and established a network of related aldose configurations that he called the D-family. The mirror images of these configurations were then designated the L-family of aldoses. To illustrate using present day knowledge, Fischer projection formulas and names for the D-aldose family (three to six-carbon atoms) are shown below, with the asymmetric carbon atoms (chiral centres) coloured red. The last chiral centre in an aldose chain (farthest from the aldehyde group) was chosen by Fischer as the D / L designator site. If the hydroxyl group in the projection formula pointed to the right, it was defined as a member of the D-family. A left directed hydroxyl group (the mirror image) then represented the L-family. Fischer's initial assignment of the D-configuration had a 50:50 chance of being right, but all his subsequent conclusions concerning the relative configurations of various aldoses were soundly based. In 1951 x-ray fluorescence studies of (+)-tartaric acid, carried out in the Netherlands by Johannes Martin Bijvoet (pronounced "buy foot"), proved that Fischer's choice was correct.

It is important to recognize that the sign of a compound's specific rotation (an experimental number) does not correlate with its configuration (D or L). It is a simple matter to measure an optical rotation with a polarimetre. Determining an absolute configuration usually requires chemical interconversion with known compounds by stereospecific reaction paths.

Important Reactions

Emil Fischer made use of several key reactions in the course of his carbohydrate studies. These are described here, together with the information that each delivers.

Oxidation

As noted above, sugars may be classified as reducing or non-reducing based on their reactivity with Tollens', Benedict's or Fehling's reagents. If a sugar is oxidized by these reagents it is called reducing, since the oxidant ($Ag^{(+)}$ or $Cu^{(+2)}$) is reduced in the reaction, as evidenced by formation of a silver mirror or precipitation of cuprous oxide. The Tollens' test is commonly used to detect aldehyde functions; and because of the facile interconversion of ketoses and aldoses under the basic conditions of this test, ketoses such as fructose also react and are classified as reducing sugars.

When the aldehyde function of an aldose is oxidized to a carboxylic acid the product is called an aldonic acid. Because of the 2° hydroxyl functions that are also present in these compounds, a mild oxidizing agent such as hypobromite must be used for this conversion (equation 1). If both ends of an aldose chain are oxidized to carboxylic acids the product is called an aldaric acid. By converting an aldose to its corresponding aldaric acid derivative, the ends of the chain become identical (this could also be accomplished by reducing the aldehyde to CH_2OH, as noted below). Such an operation will disclose any latent symmetry in the remaining molecule. Thus, ribose, xylose, allose and galactose yield achiral aldaric acids which are, of course, not optically active. The ribose oxidation is shown in equation 2 below. Other aldose sugars may give identical chiral aldaric acid products, implying a unique configurational relationship. The examples of arabinose and lyxose shown in equation 3 above illustrate this result. Remember, a Fischer projection formula may be rotated by 180° in the plane of projection without changing its configuration.

Reduction

Sodium borohydride reduction of an aldose makes the ends of the resulting alditol chain identical, $HOCH_2(CHOH)_nCH_2OH$, thereby accomplishing the same configurational change produced by oxidation to an aldaric acid. Thus, allitol and galactitol from reduction of allose and galactose are achiral, and altrose and talose are reduced to the same chiral alditol. A summary of these redox reactions, and derivative nomenclature is given in the following table.

Derivatives of $HOCH_2(CHOH)_nCHO$		
HOBr Oxidation	——>	$HOCH_2(CHOH)_nCO_2H$ an Aldonic Acid
HNO_3 Oxidation	——>	$H_2OC(CHOH)_nCO_2H$ an Aldaric Acid
$NaBH_4$ Reduction	——>	$HOCH_2(CHOH)_nCH_2OH$ an Alditol

Osazone Formation

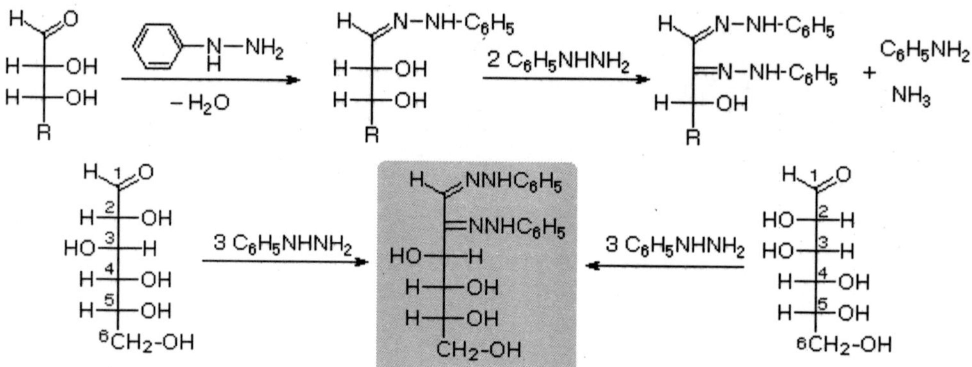

D-(+)-glucose common D-osazone D-(+)-mannose

The osazone reaction was developed and used by Emil Fischer to identify aldose sugars differing in configuration only at the alpha-carbon. The upper equation shows the general form of the osazone reaction, which effects an alpha-carbon oxidation with formation of a bis-phenylhydrazone, known as an osazone. Application of the osazone reaction to D-glucose and D-mannose demonstrates that these compounds differ in configuration only at C-2.

Chain Shortening and Lengthening

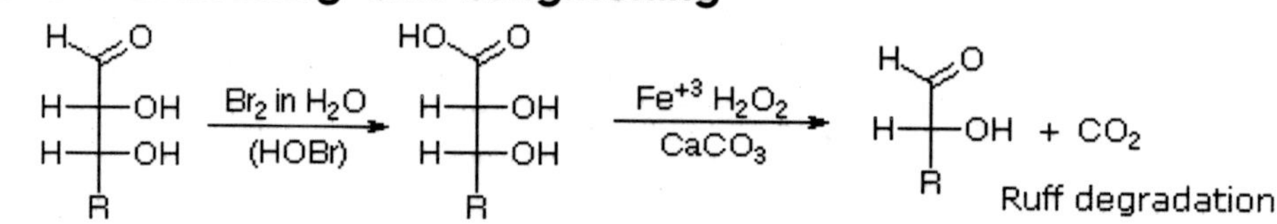

C_n aldose C_n aldonic acid C_{n-1} aldose

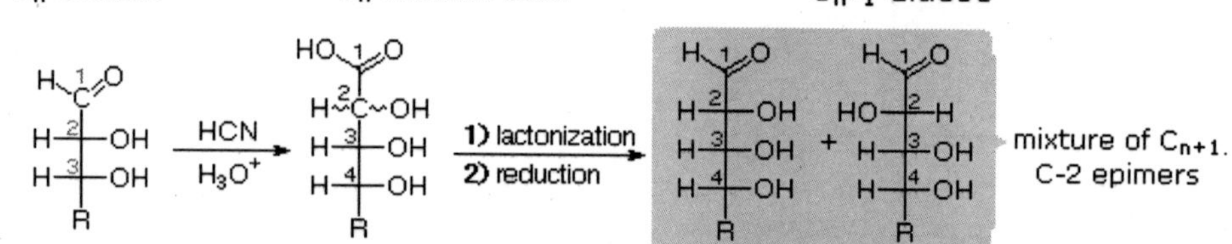

C_n aldose C_{n+1} aldonic acid Kiliani-Fischer synthesis

These two procedures permit an aldose of a given size to be related to homologous smaller and larger aldoses. The importance of these relationships may be seen in the array of aldose structures presented earlier, where the structural connections are given by the dashed blue lines. Thus Ruff degradation of the pentose arabinose gives the tetrose erythrose. Working in the opposite direction, a Kiliani-Fischer synthesis applied to arabinose gives a mixture of glucose and mannose. An alternative chain shortening procedure known as the Wohl degradation is essentially the reverse of the Kiliani-Fischer synthesis.

Using these reactions we can now follow Fischer's train of logic in assigning the configuration of D-glucose.

1. Ribose and arabinose (two well known pentoses) both gave erythrose on Ruff degradation. As expected, Kiliani-Fischer synthesis applied to erythrose gave a mixture of ribose and arabinose.
2. Oxidation of erythrose gave an achiral (optically inactive) aldaric acid. This defines the configuration of erythrose.
3. Oxidation of ribose gave an achiral (optically inactive) aldaric acid. This defines the configuration of both ribose and arabinose.
4. Ruff shortening of glucose gave arabinose, and Kiliani-Fischer synthesis applied to arabinose gave a mixture of glucose and mannose.
5. Glucose and mannose are therefore epimers at C-2, a fact confirmed by the common product from their osazone reactions.
6. A pair of structures for these epimers can be written, but which is glucose and which is mannose?

In order to determine which of these epimers was glucose, Fischer made use of the inherent C_2 symmetry in the four-carbon dissymmetric core of one epimer (B). This is shown in the following diagram by a red dot where the symmetry axis passes through the projection formula. Because of this symmetry, if the aldehyde and 1°-alcohol functions at the ends of the chain are exchanged, epimer B would be unchanged; whereas A would be converted to a different compound. By clicking on the diagram, the consequences of such an exchange will be displayed.

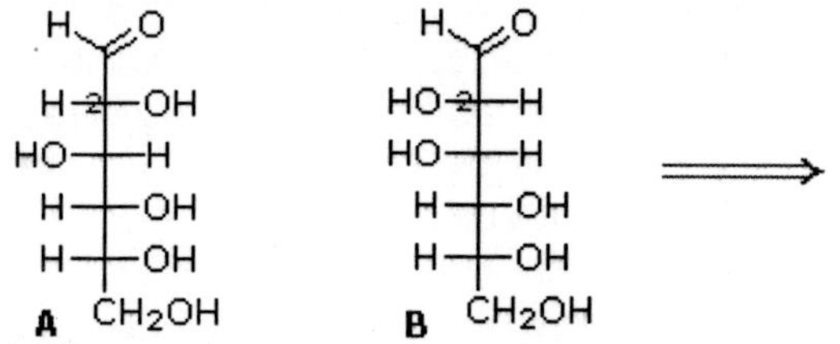

A pair of C-2 epimers.
One is glucose, the other is mannose.

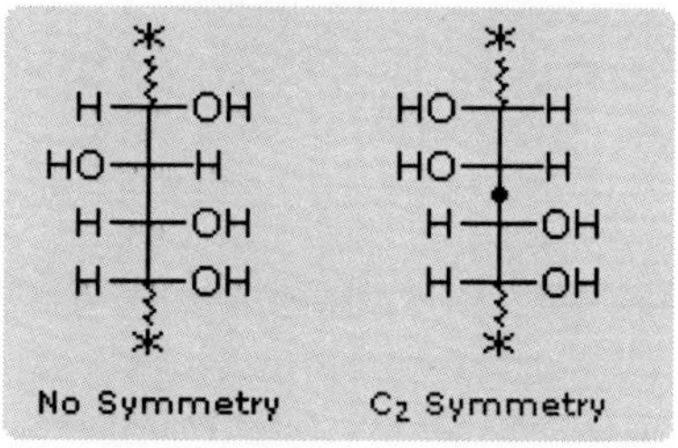

Four-Carbon Dissymmetric Core

Fischer looked for and discovered a second aldohexose that represented the end group exchange for the epimer lacking the latent C_2 symmetry (A). This compound was L-(+)-gulose, and its exchange relationship to D-(+)-glucose was demonstrated by oxidation to a common aldaric acid product. The remaining epimer is therefore mannose.

Ketoses

If a monosaccharide has a carbonyl function on one of the inner atoms of the carbon chain it is classified as a ketose. Dihydroxyacetone may not be a sugar, but it is included as the ketose analog of glyceraldehyde. The carbonyl group is commonly found at C-2, as illustrated by the following examples (chiral centres are coloured red). As expected, the carbonyl function of a ketose may be reduced by sodium borohydride, usually to a mixture of epimeric products. D-Fructose, the sweetest of the common natural sugars, is for example reduced to a mixture of D-glucitol (sorbitol) and D-mannitol, named after the aldohexoses from which they may also be obtained by analogous reduction. Mannitol is itself a common natural carbohydrate.

Although the ketoses are distinct isomers of the aldose monosaccharides, the chemistry of both classes is linked due to their facile interconversion in the presence of acid or base catalysts. This interconversion, and the corresponding epimerization at sites alpha to the carbonyl functions, occurs by way of an enediol tautomeric intermediate.

Ketose Examples

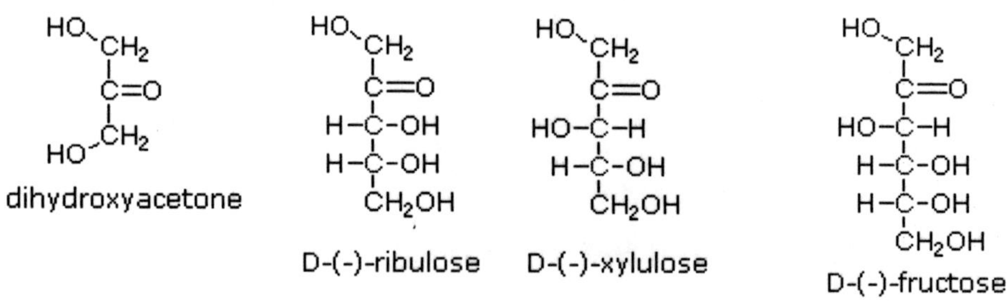

Because of base-catalyzed isomerizations of this kind, the Tollens' reagent is not useful for distinguishing aldoses from ketoses or for specific oxidation of aldoses to the corresponding aldonic acids. Oxidation by HOBr is preferred for the latter conversion.

Anomeric Forms of Glucose

Fischer's brilliant elucidation of the configuration of glucose did not remove all uncertainty concerning its structure. Two different crystalline forms of glucose were reported in 1895. Each of these gave all the characteristic reactions of glucose, and when dissolved in water equilibrated to the same mixture. This equilibration takes place over a period of many minutes, and the change in optical activity that occurs is called mutarotation. These facts are summarized in the diagram below.

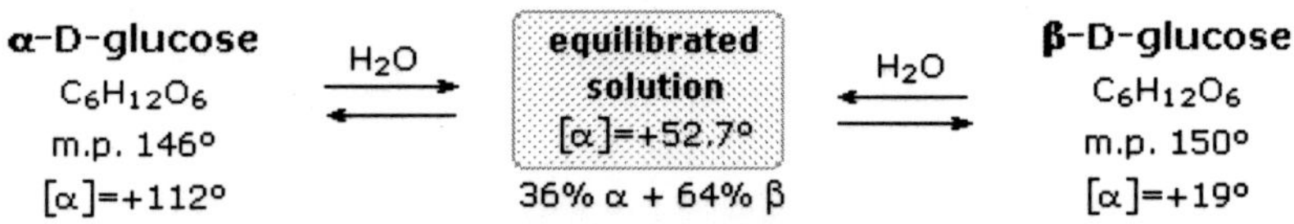

When glucose was converted to its pentamethyl ether (reaction with excess CH_3I & AgOH), two different isomers were isolated, and neither exhibited the expected aldehyde reactions. Acid-catalyzed hydrolysis of the pentamethyl ether derivatives, however, gave a tetramethyl derivative that was oxidized by Tollen's reagent and reduced by sodium borohydride, as expected for an aldehyde.

The search for scientific truth often proceeds in stages, and the structural elucidation of glucose serves as a good example. It should be clear from the new evidence presented above, that the open chain pentahydroxyhexanal structure drawn above must be modified. Somehow a new stereogenic centre must be created, and the aldehyde must be deactivated in the pentamethyl derivative. A simple solution to this dilemma is achieved by converting the open aldehyde structure for glucose into a cyclic hemiacetal, called a glucopyranose, as shown in the following diagram. The linear aldehyde is tipped on its side, and rotation about the C4-C5 bond brings the C5-hydroxyl function close to the aldehyde carbon. For ease of viewing, the six-membered hemiacetal structure is drawn as a flat hexagon, but it actually assumes a chair conformation. The hemiacetal carbon atom (C-1) becomes a new stereogenic centre, commonly referred to as the anomeric carbon, and the α and β-isomers are called anomers.

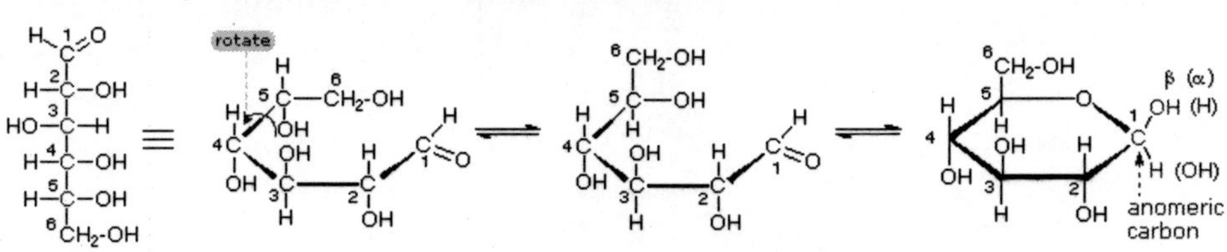

D-(+)-glucose D-glucopyranose

We can now consider how this modification of the glucose structure accounts for the puzzling facts noted above. First, we know that hemiacetals are in equilibrium with their carbonyl and alcohol components when in solution. Consequently, fresh solutions of either alpha or beta-glucose crystals in water should establish an equilibrium mixture of both anomers, plus the open chain chain form. Note that despite the very low concentration of the open chain aldehyde in this mixture, typical chemical reactions of aldehydes take place rapidly.

Second, a pentamethyl ether derivative of the pyranose structure converts the hemiacetal function to an acetal. Acetals are stable to base, so this product should not react with Tollen's reagent or be reduced by sodium borohydride. Acid hydrolysis of acetals regenerates the carbonyl and alcohol components, and in the case of the glucose derivative this will be a tetramethyl ether of the pyranose hemiacetal. This compound will, of course, undergo typical aldehyde reactions.

Cyclic Forms of Monosaccharides

As noted above, the preferred structural form of many monosaccharides may be that of a cyclic hemiacetal. Five and six-membered rings are favoured over other ring sizes because of their low angle and eclipsing strain. Cyclic structures of this kind are termed furanose (five-membered) or pyranose (six-membered), reflecting the ring size relationship to the common heterocyclic compounds furan and pyran shown on the right. Ribose, an important aldopentose, commonly adopts a furanose structure, as shown in the following illustration. By convention for the D-family, the five-membered furanose ring is drawn in an edgewise projection with the ring oxygen positioned away from the viewer. The anomeric carbon atom (coloured red here) is placed on the right. The upper bond to this carbon is defined as beta, the lower bond then is alpha.

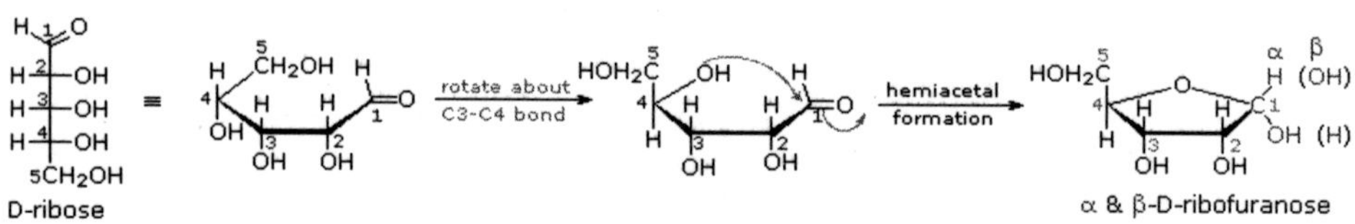

The cyclic pyranose forms of various monosaccharides are often drawn in a flat projection known as a Haworth formula, after the British chemist, Norman Haworth. As with the furanose ring, the anomeric carbon is placed on the right with the ring oxygen to the back of the edgewise view. In the D-family, the alpha and beta bonds have the same orientation defined for the furanose ring (beta is up & alpha is down). These Haworth formulas are convenient for displaying stereochemical relationships, but do not represent the true shape of the molecules. We know that these molecules are actually puckered in a fashion we call a chair conformation. Examples of four typical pyranose structures are shown below, both as Haworth projections and as the more representative chair conformers. The anomeric carbons are coloured red.

Examples of Some Pyranose Forms of Hexoses

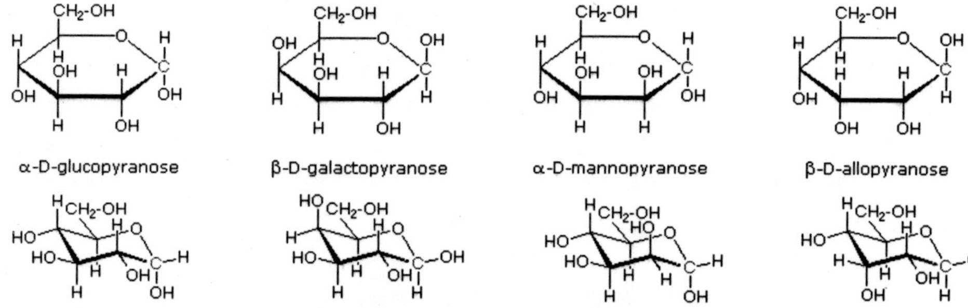

Models of these glucose, galactose, mannose and allose pyranose structures may be viewed by Clicking Here.

A practice page for examining the configurations of aldohexoses may be viewed by Clicking Here.

The size of the cyclic hemiacetal ring adopted by a given sugar is not constant, but may vary with substituents and other structural features. Aldolhexoses usually form pyranose rings and their pentose homologs tend to prefer the furanose form, but there are many counter examples. The formation of acetal derivatives illustrates how subtle changes may alter this selectivity. A pyranose structure for D-glucose is drawn in the rose-shaded box on the left. Acetal derivatives have been prepared by acid-catalyzed reactions with benzaldehyde and acetone. As a rule, benzaldehyde forms six-membered cyclic acetals, whereas acetone prefers to form five-membered acetals. The top equation shows the formation and some reactions of the 4,6-O-benzylidene acetal, a commonly employed protective group. A methyl glycoside derivative of this compound leaves the C-2 and C-3 hydroxyl groups exposed to reactions such as the periodic acid cleavage, shown as the last step. The formation of an isopropylidene acetal at C-1 and C-2, centre structure, leaves the C-3 hydroxyl as the only unprotected function. Selective oxidation to a ketone is then possible. Finally, direct di-O-isopropylidene derivatization of glucose by reaction with excess acetone results in a change to a furanose structure in which the C-3 hydroxyl is again unprotected. However, the same reaction with D-galactose, shown in the blue-shaded box, produces a pyranose product in which the C-6 hydroxyl is unprotected. Both derivatives do not react with Tollens' reagent. This difference in behaviour is attributed to the cis-orientation of the C-3 and C-4 hydroxyl groups in galactose, which permits formation of a less strained five-membered cyclic acetal, compared with the trans-C-3 and C-4 hydroxyl groups in glucose. Derivatizations of this kind permit selective reactions to be conducted at different locations in these highly functionalized molecules.

Glycosides

Acetal derivatives formed when a monosaccharide reacts with an alcohol in the presence of an acid catalyst are called glycosides. This reaction is illustrated for glucose and methanol in the diagram below. In naming of glycosides, the "ose" suffix of the sugar name is replaced by "oside", and the alcohol group name is placed first. As is generally true for most acetals, glycoside formation involves the loss of an equivalent of water. The diether product is stable to base and alkaline oxidants such as Tollen's reagent. Since acid-catalyzed aldolization is reversible, glycosides may be hydrolyzed back to their alcohol and sugar components by aqueous acid.

The anomeric methyl glucosides are formed in an equilibrium ratio of 66% alpha to 34% beta. From the structures in the previous diagram, we see that pyranose rings prefer chair conformations in which the largest number of substituents are equatorial. In the case of glucose, the substituents on the beta-anomer are all equatorial, whereas the C-1 substituent in the alpha-anomer changes to axial. Since substituents on cyclohexane rings prefer an equatorial location over axial (methoxycyclohexane is 75% equatorial), the preference for alpha-glycopyranoside formation is unexpected, and is referred to as the anomeric effect.

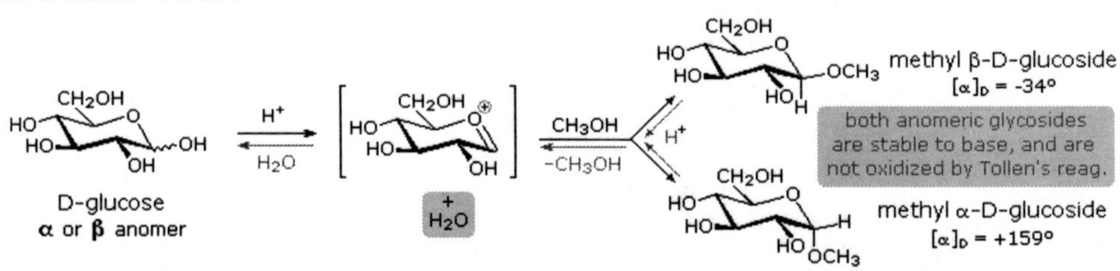

Glycosides abound in biological systems. By attaching a sugar moiety to a lipid or benzenoid structure, the solubility and other properties of the compound may be changed substantially. Because of the important modifying influence of such derivatization, numerous enzyme systems, known as glycosidases, have evolved for the attachment and removal of sugars from alcohols, phenols and amines. Chemists refer to the sugar component of natural glycosides as the glycon and the alcohol component as the aglycon. Salicin, one of the oldest herbal remedies known, was the model for the synthetic analgesic aspirin. A large class of hydroxylated, aromatic oxonium cations called anthocyanins provide the red, purple and blue colours of many flowers, fruits and some vegetables. Peonin is one example of this class of natural pigments, which exhibit a pronounced pH colour dependence. The oxonium moiety is only stable in acidic environments, and the colour changes or disappears when base is added. The complex changes that occur when wine is fermented and stored are in part associated with glycosides of anthocyanins. Finally, amino derivatives of ribose, such as cytidine play important roles in biological phosphorylating agents, coenzymes and information transport and storage materials.

Disaccharides

When the alcohol component of a glycoside is provided by a hydroxyl function on another monosaccharide, the compound is called a disaccharide. Four examples of disaccharides composed of two glucose units are shown in the following diagram. The individual glucopyranose rings are labelled A and B, and the glycoside bonding is circled in light blue. Notice that the glycoside bond may be alpha, as in maltose and trehalose, or beta as in cellobiose and gentiobiose. Acid-catalyzed hydrolysis of these disaccharides yields glucose as the only product. Enzyme-catalyzed hydrolysis is selective for a specific glycoside bond, so an alpha-glycosidase cleaves maltose and trehalose to glucose, but does not cleave cellobiose or gentiobiose. A beta-glycosidase has the opposite activity.

In order to draw a representative structure for cellobiose, one of the glucopyranose rings must be rotated by 180°, but this feature is often omitted in favour of retaining the usual perspective for the individual rings. The bonding between the glucopyranose rings in cellobiose and maltose is from the anomeric carbon in ring A to the C-4 hydroxyl group on ring B. This leaves the anomeric carbon in ring B free, so cellobiose and maltose both may assume alpha and beta anomers at that site (the beta form is shown in the diagram). Gentiobiose has a beta-glycoside link, originating at C-1 in ring A and terminating at C-6 in ring B. Its alpha-anomer is drawn in the diagram. Because cellobiose, maltose and gentiobiose are hemiacetals they are all reducing sugars (oxidized by Tollen's reagent). Trehalose, a disaccharide found in certain mushrooms, is a bis-acetal, and is therefore a non-reducing sugar. A systematic nomenclature for disaccharides exists, but as the following examples illustrate, these are often lengthy.

Cellobiose : 4-O-β-D-Glucopyranosyl-D-glucose (the beta-anomer is drawn)

Maltose : 4-O-α-D-Glucopyranosyl-D-glucose (the beta-anomer is drawn)

Gentiobiose : 6-O-β-D-Glucopyranosyl-D-glucose (the alpha-anomer is drawn)

Trehalose : α-D-Glucopyranosyl-α-D-glucopyranoside

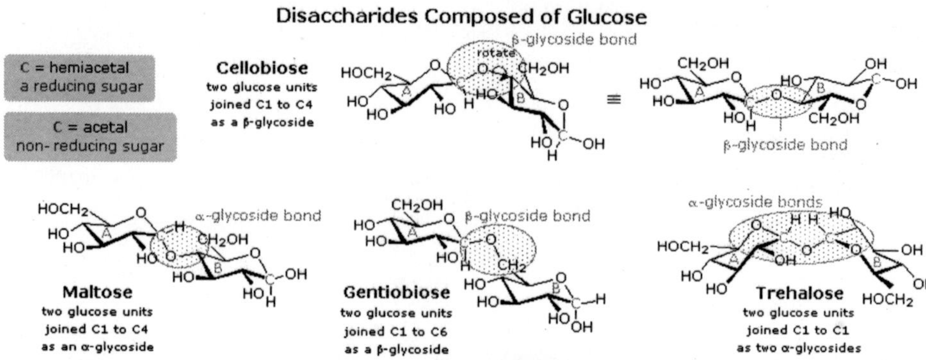

Although all the disaccharides shown here are made up of two glucopyranose rings, their properties differ in interesting ways. Maltose, sometimes called malt sugar, comes from the hydrolysis of starch. It is about one third as sweet as cane sugar (sucrose), is easily digested by humans, and is fermented by yeast. Cellobiose is obtained by the hydrolysis of cellulose. It has virtually no taste, is indigestible by humans, and is not fermented by yeast. Some bacteria have beta-glucosidase enzymes that hydrolyze the glycosidic bonds in cellobiose and cellulose. The presence of such bacteria in the digestive tracts of cows and termites permits these animals to use cellulose as a food. Finally, it may be noted that trehalose has a distinctly sweet taste, but gentiobiose is bitter.

Disaccharides made up of other sugars are known, but glucose is often one of the components. Two important examples of such mixed disaccharides will be displayed above by clicking on the diagram. Lactose, also known as milk sugar, is a galactose-glucose compound joined as a beta-glycoside. It is a reducing sugar because of the hemiacetal function remaining in the glucose moiety. Many adults, particularly those from regions where milk is not a dietary staple, have a metabolic intolerance for lactose. Infants have a digestive enzyme which cleaves the beta-glycoside bond in lactose, but production of this enzyme stops with weaning. Cheese is less subject to the lactose intolerance problem, since most of the lactose is removed with the whey. Sucrose, or cane sugar, is our most commonly used sweetening agent. It is a non-reducing disaccharide composed of glucose and fructose joined at the anomeric carbon of each by glycoside bonds (one alpha and one beta). In the formula shown here the fructose ring has been rotated 180° from its conventional perspective.

Polysaccharides

As the name implies, polysaccharides are large high-molecular weight molecules constructed by joining monosaccharide units together by glycosidic bonds. They are sometimes called glycans. The most important compounds in this class, cellulose, starch and glycogen are all polymers of glucose. This is easily demonstrated by acid-catalyzed hydrolysis to the monosaccharide. Since partial hydrolysis of cellulose gives varying amounts of cellobiose, we conclude the glucose units in this macromolecule are joined by beta-glycoside bonds between C-1 and C-4 sites of adjacent sugars. Partial hydrolysis of starch and glycogen produces the disaccharide maltose together with low molecular weight dextrans, polysaccharides in which glucose molecules are joined by alpha-glycoside links between C-1 and C-6, as well as the alpha C-1 to C-4 links found in maltose. Polysaccharides built from other monosaccharides (e.g. mannose, galactose, xylose and arabinose) are also known, but will not be discussed here.

Over half of the total organic carbon in the earth's biosphere is in cellulose. Cotton fibres are essentially pure cellulose, and the wood of bushes and trees is about 50% cellulose. As a polymer of glucose, cellulose has the formula $(C_6H_{10}O_5)_n$ where n ranges from 500 to 5,000, depending on the source of the polymer. The glucose units in cellulose are linked in a linear fashion, as shown in the drawing below. The beta-glycoside bonds permit these chains to stretch out, and this conformation is stabilized by intramolecular hydrogen bonds. A parallel orientation of adjacent chains is also favoured by intermolecular hydrogen bonds. Although an individual hydrogen bond is relatively weak, many such bonds acting together can impart great stability to certain conformations of large molecules. Most animals cannot digest cellulose as a food, and in the diets of humans this part of our vegetable intake functions as roughage and is eliminated largely unchanged. Some animals (the cow and termites, for example) harbour intestinal microorganisms that breakdown cellulose into monosaccharide nutrients by the use of beta-glycosidase enzymes.

Cellulose is commonly accompanied by a lower molecular weight, branched, amorphous polymer called hemicellulose. In contrast to cellulose, hemicellulose is structurally weak and is easily hydrolyzed by dilute acid or base. Also, many enzymes catalyze its hydrolysis. Hemicelluloses are composed of many D-pentose sugars, with xylose being the major component. Mannose and mannuronic acid are often present, as well as galactose and galacturonic acid.

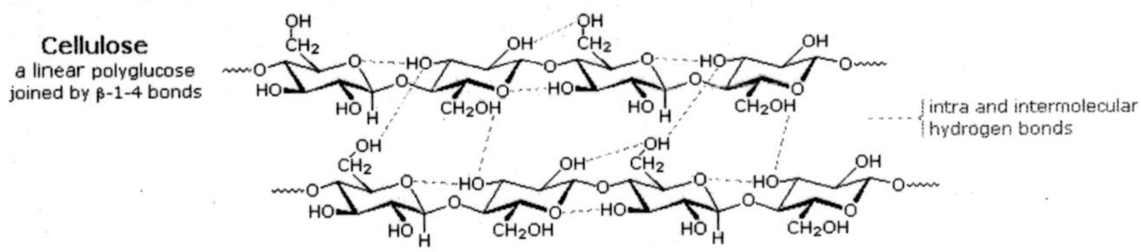

Starch is a polymer of glucose, found in roots, rhizomes, seeds, stems, tubers and corms of plants, as microscopic granules having characteristic shapes and sizes. Most animals, including humans, depend on these plant starches for nourishment. The structure of starch is more complex than that of cellulose. The intact granules are insoluble in cold water, but grinding or swelling them in warm water causes them to burst.

The released starch consists of two fractions. About 20% is a water soluble material called amylose. Molecules of amylose are linear chains of several thousand glucose units joined by alpha C-1 to C-4 glycoside bonds. Amylose solutions are actually dispersions of hydrated helical micelles. The majority of the starch is a much higher molecular weight substance, consisting of nearly a million glucose units, and called amylopectin. Molecules of amylopectin are branched networks built from C-1 to C-4 and C-1 to C-6 glycoside links, and are essentially water insoluble. The branching in this diagram is exaggerated, since on average, branches only occur every twenty five glucose units. Hydrolysis of starch, usually by enzymatic reactions, produces a syrupy liquid consisting largely of glucose. When cornstarch is the feedstock, this product is known as corn syrup. It is widely used to soften texture, add volume, prohibit crystallization and enhance the flavour of foods.

Glycogen is the glucose storage polymer used by animals. It has a structure similar to amylopectin, but is even more highly branched (about every tenth glucose unit). The degree of branching in these polysaccharides may be measured by enzymatic or chemical analysis.

Synthetic Modification of Cellulose

Cotton, probably the most useful natural fibre, is nearly pure cellulose. The manufacture of textiles from cotton involves physical manipulation of the raw material by carding, combing and spinning selected fibres. For fabrics the best cotton has long fibres, and short fibres or cotton dust are removed. Crude cellulose is also available from wood pulp by dissolving the lignan matrix surrounding it. These less desirable cellulose sources are widely used for making paper. In order to expand the ways in which cellulose can be put to practical use, chemists have devised techniques for preparing solutions of cellulose derivatives that can be spun into fibres, spread into a film or cast in various solid forms. A key factor in these transformations are the three free hydroxyl groups on each glucose unit in the cellulose chain, —$[C_6H_7O(OH)_3]_n$—. Esterification of these functions leads to polymeric products having very different properties compared with cellulose itself.

Cellulose Nitrate, first prepared over 150 years ago by treating cellulose with nitric acid, is the earliest synthetic polymer to see general use. The fully nitrated compound, —$[C_6H_7O(ONO_2)_3]_n$—, called guncotton, is explosively flammable and is a component of smokeless powder. Partially nitrated cellulose is called pyroxylin. Pyroxylin is soluble in ether and at one time was used for photographic film and lacquers. The high flammability of pyroxylin caused many tragic cinema fires during its period of use. Furthermore, slow hydrolysis of pyroxylin yields nitric acid, a process that contributes to the deterioration of early motion picture films in storage.

Cellulose Acetate, —$[C_6H_7O(OAc)_3]_n$—, is less flammable than pyroxylin, and has replaced it in most applications. It is prepared by reaction of cellulose with acetic anhydride and an acid catalyst. The properties of the product vary with the degree of acetylation. Some chain shortening occurs unavoidably in the preparations. An acetone solution of cellulose acetate may be forced through a spinneret to generate filaments, called acetate rayon, that can be woven into fabrics.

Viscose Rayon, is prepared by formation of an alkali soluble xanthate derivative that can be spun into a fibre that reforms the cellulose polymer by acid quenching. The following general equation illustrates these transformations. The product fibre is called viscose rayon.

ROH	NaOH	$RO^{(-)}Na^{(+)}$ + S=C=S		$RO\text{-}CS_2^{(-)}Na^{(+)}$	$H_3O^{(+)}$	ROH
cellulose				viscose solution		rayon

Many complex polysaccharides are found in nature. The galactomannans, consisting of a mannose backbone with galactose side groups, are an interesting and useful example. A (1-4)-linked beta-D-mannose chain is adorned with 1-6-linked alpha-D-galactose units, as shown in the diagram below. The ratio of galactose to mannose usually ranges from 1:2 to 1:4. An important source of this substance is the guar bean, grown principally in northwestern India, and Pakistan. Altough guar protein is not of nutritional value to humans (guar means 'cow food' in Hindi), the bean is important as a source of guar gum, a galactomannan which forms a gel in water. The food industry uses this material as a stabilizer in ice cream, cream cheese and salad dressings.

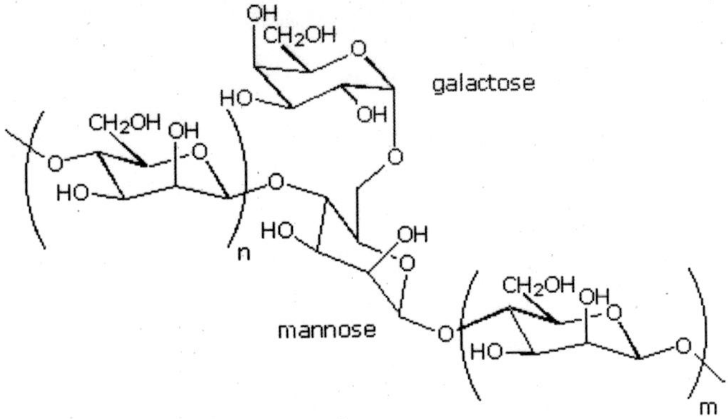

A Galactomannan

Recently, guar gum has becomes an essential ingredient for mining oil and natural gas in a process called hydraulic fracturing. The demand for guar has increased to such a degree that the poor farmers in northwestern India have had their lives transformed by unexpected wealth.

Nucleic Acids

With millions of different bacterial species living in soil, the microbial community is extremely complex, varying at very small scales. Microbe-driven functions are essential for most processes in soil. Thus, a better understanding of this microbial diversity will be invaluable for the management of the various soil functions.

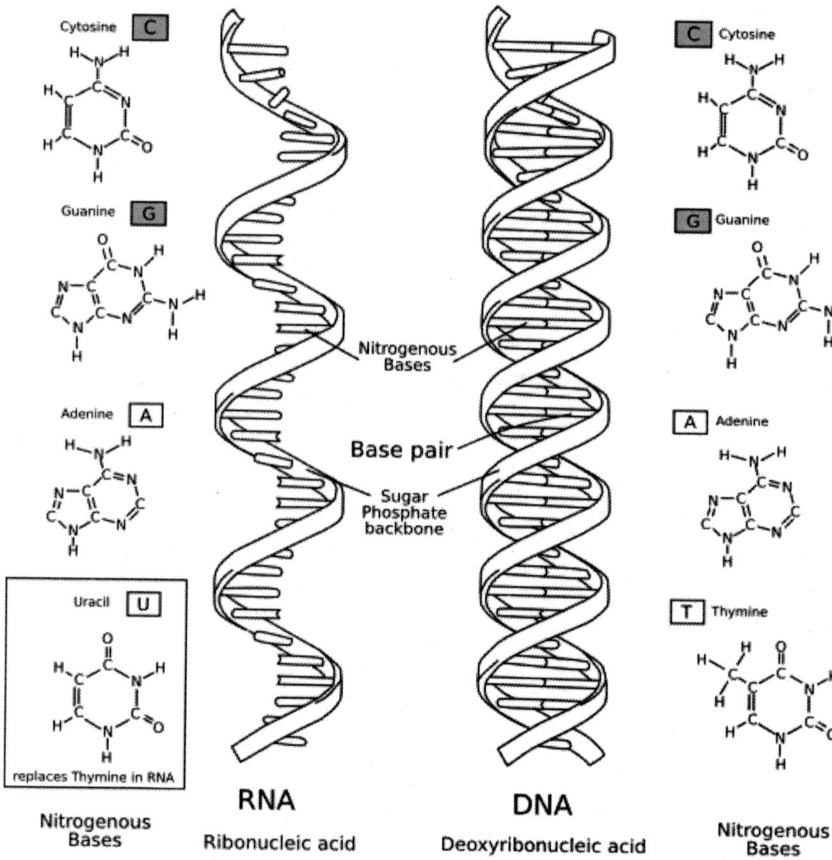

Figure: *A comparison of the two principal nucleic acids: RNA (left) and DNA (right), showing the helices and nucleobases each employs.*

Nucleic acids are polymeric macromolecules, or large biological molecules, essential for all known forms of life. Nucleic acids, which include DNA (deoxyribonucleic acid) and RNA (ribonucleic acid), are made from monomers known as nucleotides. Each nucleotide has three components: a 5-carbon sugar, a phosphate group, and a nitrogenous base. If the sugar is deoxyribose, the polymer is DNA. If the sugar is ribose, the polymer is RNA.

Together with proteins, nucleic acids are the most important biological macromolecules; each is found in abundance in all living things, where they function in encoding, transmitting and expressing genetic information—in other words, information is conveyed through the nucleic acid sequence, or the order of nucleotides within a DNA or RNA molecule. Strings of nucleotides strung together in a specific sequence are the mechanism for storing and transmitting hereditary, or genetic, information via protein synthesis. Nucleic acids were discovered by Friedrich Miescher in 1869. Experimental studies of nucleic acids constitute a major part of modern biological and medical research, and form a foundation for genome and forensic science, as well as the biotechnology and pharmaceutical industries.

Occurrence and Nomenclature

The term *nucleic acid* is the overall name for DNA and RNA, members of a family of biopolymers, and is synonymous with *polynucleotide*. Nucleic acids were named for their initial discovery within the nucleus, and for the presence of phosphate groups (related to phosphoric acid). Although first discovered within the nucleus of eukaryotic cells, nucleic acids are now known to be found in all life forms as well as some nonliving entities, including within bacteria, archaea, mitochondria, chloroplasts, viruses and viroids. All living cells contain both DNA and RNA (except some cells such as mature red blood cells), while viruses contain either DNA or RNA, but usually not both. The basic component of biological nucleic acids is the nucleotide, each of which contains a pentose sugar (ribose or deoxyribose), a phosphate group, and a nucleobase. Nucleic acids are also generated within the laboratory, through the use of enzymes (DNA and RNA polymerases) and by solid-phase chemical synthesis. The chemical methods also enable the generation of altered nucleic acids that are not found in nature, for example peptide nucleic acids.

Molecular Composition and Size

Nucleic acids are generally very large molecules. Indeed, DNA molecules are probably the largest individual molecules known. Well-studied biological nucleic acid molecules range in size from 21 nucleotides (small interfering RNA) to large chromosomes (human chromosome 1 is a single molecule that contains 247 million base pairs).

In most cases, naturally occurring DNA molecules are double-stranded and RNA molecules are single-stranded. There are numerous exceptions, however—some viruses have genomes made of double-stranded RNA and other viruses have single-stranded DNA genomes, and, in some circumstances, nucleic acid structures with three or four strands can form.

Nucleic acids are linear polymers (chains) of nucleotides. Each nucleotide consists of three components: a purine or pyrimidine nucleobase (sometimes termed *nitrogenous base* or simply *base*), a pentose sugar, and a phosphate group. The substructure consisting of a nucleobase plus sugar is termed a nucleoside. Nucleic acid types differ in the structure of the sugar in their nucleotides - DNA contains 2'-deoxyribose while RNA contains ribose (where the only difference is the presence of a hydroxyl group). Also, the nucleobases found in the two nucleic acid types are different: adenine, cytosine, and guanine are found in both RNA and DNA, while thymine occurs in DNA and uracil occurs in RNA.

The sugars and phosphates in nucleic acids are connected to each other in an alternating chain (sugar-phosphate backbone) through phosphodiester linkages. In conventional nomenclature, the carbons to which the phosphate groups attach are the 3'-end and the 5'-end carbons of the sugar. This gives nucleic acids directionality, and the ends of nucleic acid molecules are referred to as 5'-end and 3'-end. The nucleobases are joined to the sugars via an N-glycosidic linkage involving a nucleobase ring nitrogen (N-1 for pyrimidines and N-9 for purines) and the 1' carbon of the pentose sugar ring.

Non-standard nucleosides are also found in both RNA and DNA and usually arise from modification of the standard nucleosides within the DNA molecule or the primary (initial) RNA transcript. Transfer RNA (tRNA) molecules contain a particularly large number of modified nucleosides.

Topology

Double-stranded nucleic acids are made up of complementary sequences, in which extensive Watson-Crick base pairing results in a highly repeated and quite uniform double-helical three-dimensional structure. In contrast, single-stranded RNA and DNA molecules are not constrained to a regular double helix, and can adopt highly complex three-dimensional structures that are based on short stretches of intramolecular base-paired sequences that include both Watson-Crick and noncanonical base pairs, as well as a wide range of complex tertiary interactions.

Nucleic acid molecules are usually unbranched, and may occur as linear and circular molecules. For example, bacterial chromosomes, plasmids, mitochondrial DNA and chloroplast DNA are usually circular double-stranded DNA molecules, while chromosomes of the eukaryotic nucleus are usually linear double-stranded DNA molecules. Most RNA molecules are linear, single-stranded molecules, but both circular and branched molecules can result from RNA splicing reactions.

Nucleic Acid Sequences

One DNA or RNA molecule differs from another primarily in the sequence of nucleotides. Nucleotide sequences are of great importance in biology since they carry the ultimate instructions that encode all biological molecules, molecular assemblies, subcellular and cellular structures, organs and organisms, and directly enable cognition, memory and behaviour. Enormous efforts have gone into the development of experimental methods to determine the nucleotide sequence of biological DNA and RNA molecules, and today hundreds of millions of nucleotides are sequenced daily at genome centres and smaller laboratories worldwide.

Types of Nucleic Acids

Deoxyribonucleic Acid: Deoxyribonucleic acid (DNA) is a nucleic acid containing the genetic instructions used in the development and functioning of all known living organisms (with the exception of RNA viruses). The DNA segments carrying this genetic information are called genes. Likewise, other DNA sequences have structural purposes, or are involved in regulating the use of this genetic information. Along with RNA and proteins, DNA is one of the three major macromolecules that are essential for all known forms of life. DNA consists of two long polymers of simple units called nucleotides, with backbones made of sugars and phosphate groups joined by ester bonds. These two strands run in opposite directions to each other and are, therefore, anti-parallel. Attached to each sugar is one of four types of molecules called nucleobases (informally, bases). It is the sequence of these four nucleobases along the backbone that encodes information. This information is read using the genetic code, which specifies the sequence of the amino acids within proteins. The code is read by copying stretches of DNA into the related nucleic acid RNA in a process called transcription. Within cells DNA is organized into long structures called chromosomes. During cell division these chromosomes are duplicated in the process of DNA replication, providing each cell its own complete set of chromosomes. Eukaryotic organisms (animals, plants, fungi, and protists) store most of their DNA inside the cell nucleus and some of their DNA in organelles, such as mitochondria or chloroplasts. In contrast, prokaryotes (bacteria and archaea) store their DNA only in the cytoplasm. Within the chromosomes, chromatin proteins such as histones compact and organize DNA. These compact structures guide the interactions between DNA and other proteins, helping control which parts of the DNA are transcribed.

Ribonucleic acid

Ribonucleic acid (RNA) functions in converting genetic information from genes into the amino acid sequences of proteins. The three universal types of RNA include transfer RNA (tRNA), messenger RNA (mRNA), and ribosomal RNA (rRNA). Messenger RNA acts to carry genetic sequence information between DNA and ribosomes, directing protein synthesis. Ribosomal RNA is a major component of the ribosome, and catalyzes peptide bond formation. Transfer RNA serves as the carrier molecule for amino acids to be used in protein synthesis, and is responsible for decoding the mRNA. In addition, many other classes of RNA are now known.

Artificial Nucleic Acid Analogs

Artificial nucleic acid analogs have been designed and synthesized by chemists, and include peptide nucleic acid, morpholino- and locked nucleic acid, as well as glycol nucleic acid and threose nucleic acid. Each of these is distinguished from naturally occurring DNA or RNA by changes to the backbone of the molecule.

Measurement of Nucleic Acid Components in Soil

Organic phosphorus was extracted from soil by digestion with 1 *N* NaOH for 19 hours at 30°C. Acidification of the alkaline extract resulted in precipitation of organic materials or humus. The soluble organic phosphorus in the acid extract was separated from excess salt and concentrated by adsorption on activated alumina. After removal from the alumina, the extract was resolved by anion-exchange chromatography. Samples of the eluants from chromatographic peaks were analyzed in a spectrophotometre. Techniques were evaluated by the addition of known quantities of nucleic acid to soils and the determination of the extent of nucleotide losses in the several steps involved.

The data showed that a Carrington soil containing 575 ppm. of organic phosphorus, and a Webster soil containing 327 ppm., contained not more than 1 ppm. and 6 ppm., respectively, of ribonucleic acid phosphorus by the above described technique. The large fraction of soil organic phosphorus was not identified, but the evidence was considered adequate that only a small portion of the soil organic phosphorus could be accounted for as nucleic acids. Purine bases were not detected in soil hydrolysates. Paper chromatograms were run on an acid hydrolysate of soil and the migration of the ultraviolet absorbing spots compared with the migration of known purine bases. Under ultraviolet light no migration of the soil hydrolysate could be detected. The absence of detectable quantities of purine bases also indicated that very little nucleic acid exists in the soil.

Phenolic Compounds in Soils

In organic chemistry, phenols, sometimes called phenolics, are a class of chemical compounds consisting of a hydroxyl group (—OH) bonded directly to an aromatic hydrocarbon group. The simplest of the class is phenol, which is also called carbolic acid C6H5OH. Phenolic compounds are classified as simple phenols or polyphenols based on the number of phenol units in the molecule.

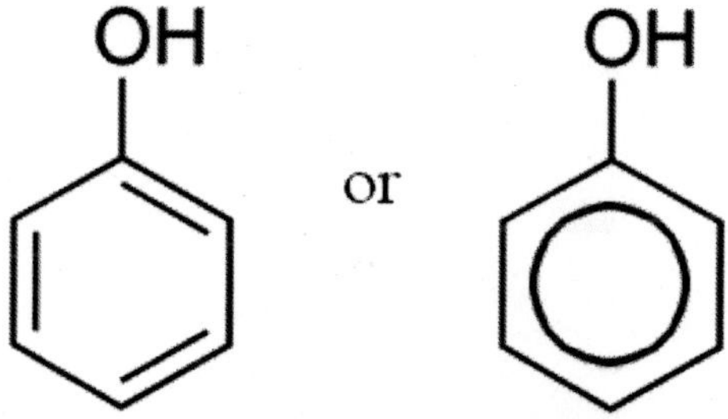

Figure: *Phenol - the simplest of the phenols.*

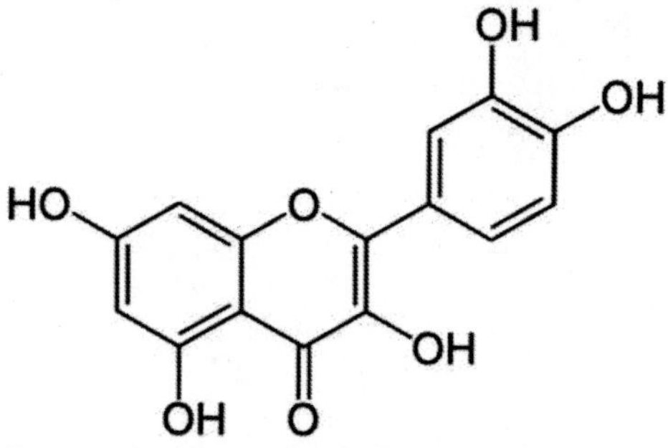

Figure: *Quercetin, a typical flavonoid, is a polyphenol*

Phenolic compounds are synthesized industrially; they also are produced by plants and microorganisms, with variation between and within species.

Although similar to alcohols, phenols have unique properties and are not classified as alcohols (since the hydroxyl group is not bonded to a *saturated* carbon atom). They have higher acidities due to the aromatic ring's tight coupling with the oxygen and a relatively loose bond between the oxygen and hydrogen. The acidity of the hydroxyl group in phenols is commonly intermediate between that of aliphatic alcohols and carboxylic acids (their pK_a is usually between 10 and 12).

Loss of a positive hydrogen ion (H^+) from the hydroxyl group of a phenol forms a corresponding negative phenolate ion or phenoxide ion, and the corresponding salts are called phenolates or phenoxides, although the term aryloxides is preferred according to the IUPAC Gold Book. Phenols can have two or more hydroxy groups bonded to the aromatic ring(s) in the same molecule. The simplest examples are the three benzenediols, each having two hydroxy groups on a benzene ring.

Organisms that synthesize phenolic compounds do so in response to ecological pressures such as pathogen and insect attack, UV radiation and wounding. As they are present in food consumed in human diets and in plants used in traditional medicine of several cultures, their role in human health and disease is a subject of research. Some phenols are germicidal and are used in formulating disinfectants. Others possess estrogenic or endocrine disrupting activity.

Chemistry

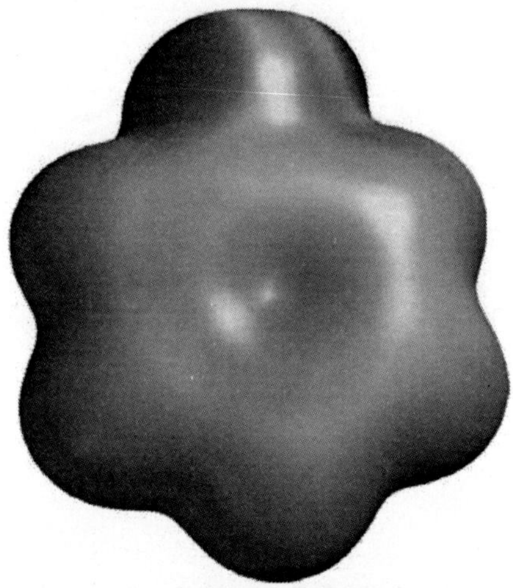

Figure: *Neutral phenol substructure "shape". An image of a computed electrostatic surface of neutral phenol, showing neutral regions in green, electronegative areas in orange-red, and the electropositive phenolic proton in blue.*

Chemical Properties: The majority of these compounds are solubles molecules but the smaller molecules can be volatiles.

Many natural phenols present chirality within their molecule. An example of such molecules is catechin. Cavicularin is an unusual macrocycle because it was the first compound isolated from nature displaying optical activity due to the presence of planar chirality and axial chirality.

Natural phenols chemically interact with many other substances. Stacking, a chemical property of molecules with aromaticity, is seen occurring between phenolic molecules. When studied in mass spectrometry, phenols easily form adduct ions with halogens. They can also interact with the food matrices or with different forms of silica (mesoporous silica, fumed silica or silica-based sol gels).

UV Visible Absorbance

Natural phenols spectral data show a typical UV absorbance characteristic of benzene aromaticity at 270 nm. However, according to Woodward's rules, bathochromic shifts often also happen suggesting the presence of delocalised π electrons arising from a conjugation between the benzene and vinyls groups. As molecules with higher conjugation levels undergo this bathochromic shift phenomenon, a part of the visible spectrum is absorbed. The wavelengths left in the process (generally in red section of the spectrum) recompose the colour of the particular substance. Acylation with cinnamic acids of anthocyanidins shifted colour tonality (CIE Lab hue angle) to purple.

Here is a series of UV visible spectra of molecules classified from left to right according to their conjugation level: The absorbance pattern responsible for the red colour of anthocyanins may be complementary to that of green chlorophyll in photosynthetically active tissues such as young *Quercus coccifera* leaves.

Oxidation

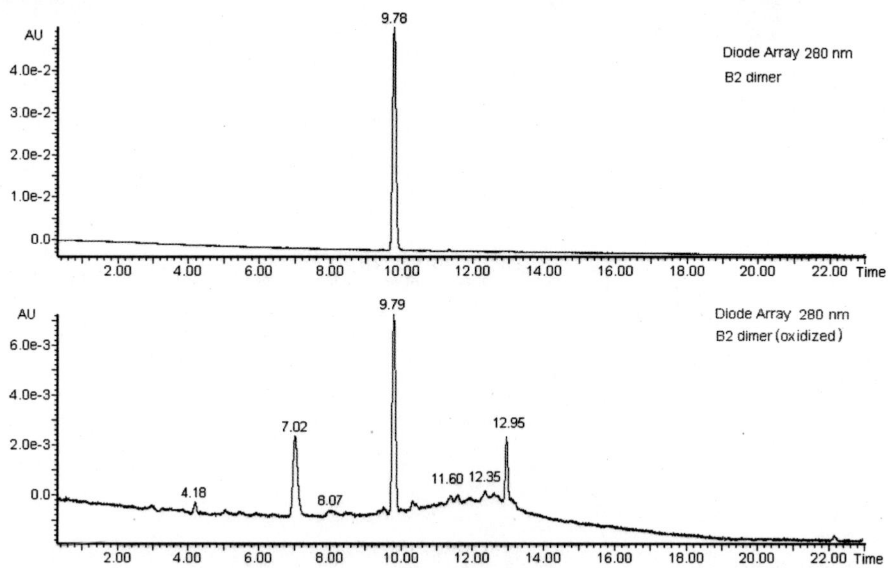

Figure: *Chromatograms showing the oxidation of a proanthocyanidin B2 dimer. New peaks have appeared in the oxidised sample.*

Natural phenols are reactive species towards oxidation, notably the complex mixture of phenolics, found in food for example, can undergo autoxidation during the ageing process. Simple natural phenols can lead to the formation of B-type procyanidins in wines or in model solutions. This is correlated to the non enzymatic browning colour change characteristic of this process. This phenomenon can be observed in foods like carrot purees. Browning associated with oxidation of phenolic compounds has also been given as the cause of cells death in calli formed in *in vitro* cultures. Those phenolics originate both from explant tissues and from explant secretions.

Phenolic Compounds

Cannabinoids	***the active constituents of cannabis***
Capsaicin	the pungent compound of chili peppers
Cresol	found in coal tar and creosote
Estradiol	estrogen - hormones
Eugenol	the main constituent of the essential oil of clove
Gallic acid	found in galls
Guaiacol	(2-methoxyphenol) - has a smokey flavour, and is found in roasted coffee, whisky, and smoke
Methyl salicylate	the major constituent of the essential oil of wintergreen
Raspberry ketone	a compound with an intense raspberry smell
Salicylic acid	precursor compound to Aspirin (chemical synthesis is used in manufacturing)
Serotonin / dopamine / adrenaline / noradrenaline	natural neurotransmitters
Thymol	(2-Isopropyl-5-methyl phenol) - found in thyme; an antiseptic that is used in mouthwashes
Tyrosine	an amino acid
Synthetic	
Phenol	the parent compound, used as a disinfectant and for chemical synthesis
Bisphenol A	and other bisphenols produced from ketones and phenol / cresol
BHT	(butylated hydroxytoluene) - a fat-soluble antioxidant and food additive
4-Nonylphenol	a breakdown product of detergents and nonoxynol-9
Orthophenyl phenol	a fungicide used for waxing citrus fruits
Picric acid	(trinitrophenol) - an explosive material
Phenolphthalein	pH indicator
Xylenol	used in antiseptics & disinfecticides

Drugs, Present and Past

Diethylstilbestrol	a synthetic estrogen with a stilbene structure; no longer marketed
L-DOPA	a dopamine prodrug used to treat Parkinson's Disease
Propofol	a short-acting intravenous anesthetic agent

Chemical properties

The majority of these compounds are soluble molecules but the smaller molecules can be volatile.

Phenols often have chiral centres. An example of such a molecule is catechin. Cavicularin is an unusual macrocycle because it was the first compound isolated from nature displaying optical activity due to the presence of planar chirality and axial chirality.

Phenols chemically interact with many other substances. Stacking, a chemical property of molecules with aromaticity, is seen occurring between phenolic molecules. When studied in mass spectrometry, phenols easily form adduct ions with halogens. They can also interact with the food matrices or with different forms of silica (mesoporous silica, fumed silica or silica-based sol gels).

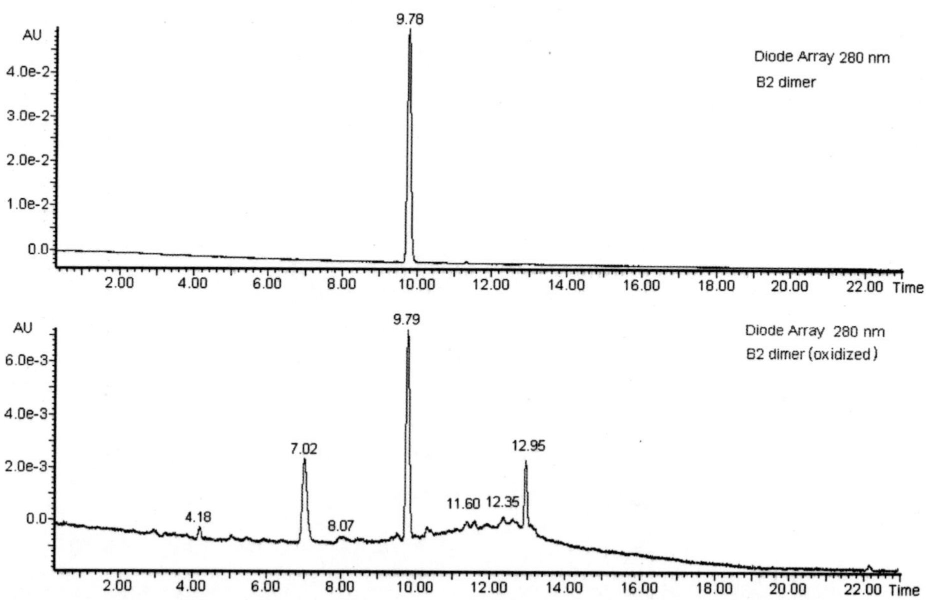

***Figure:** Chromatograms showing the oxidation of a proanthocyanidin B2 dimer. New peaks have appeared in the oxidised sample.*

Phenols are reactive species towards oxidation, notably the complex mixture of phenolics, found in food for example, can undergo autoxidation during the ageing process. Simple natural phenols can lead to the formation of B-type procyanidins in wines or in model solutions. This is correlated to the non enzymatic browning colour change characteristic of this process. This phenomenon can be observed in foods like carrot purees.

Biosynthesis

Phenolics are formed by three different biosynthetic pathways: (i) the shikimate/chorizmate or succinylbenzoate pathway, which produces the phenyl propanoid derivatives (C6–C3); (ii) the acetate/malonate or polyketide pathway, which produces the side-chain-elongated phenyl propanoids, including the large group of flavonoids (C6–C3–C6) and some quinones; and (iii) the acetate/mevalonate pathway, which produces the aromatic terpenoids, mostly monoterpenes, by dehydrogenation reactions. The aromatic amino acid phenylalanine, synthesized in the shikimic acid pathway, is the common precursor of phenol containing amino acids and phenolic compounds.

In plants, the phenolic units are esterified or methylated and are submitted to conjugation, which means that the natural phenols are mostly found in the glycoside form instead of the aglycone form.

In olive oil, tyrosol forms esters with fatty acids. In rye, alkylresorcinols are phenolic lipids.

Some acetylations involve terpenes like geraniol. Those molecules are called meroterpenes (a chemical compound having a partial terpenoid structure).

Methylations can occur by the formation of an ether bond on hydroxyl groups forming O-methylated polyphenols. In the case of the O-methylated flavone tangeritin, all of the five hydroxyls are methylated, leaving no free hydroxyls of the phenol group. Methylations can also occur on directly on a carbon of the benzene ring like in the case of poriol, a C-methylated flavonoid.

Synthesis of Phenols

Several laboratory methods for the synthesis of phenols:

- by an ester rearrangement in the Fries rearrangement
- by a rearrangement of N-phenylhydroxylamines in the Bamberger rearrangement
- by hydrolysis of phenolic esters or ethers
- by reduction of quinones
- by replacement of an aromatic amine by an hydroxyl group with water and sodium bisulfide in the Bucherer reaction

- by hydrolysis of diazonium salts
- by oligomerisation with formaldehyde + base catalysed reaction with epichlorohydrin to epoxi resin components
- by reaction with acetone/ketones to e.g. Bisphenol A, an important monomer for resins, e.g. polycarbonate (PC), epoxi resins
- by a rearrangement reaction of dienones in the dienone phenol rearrangement:

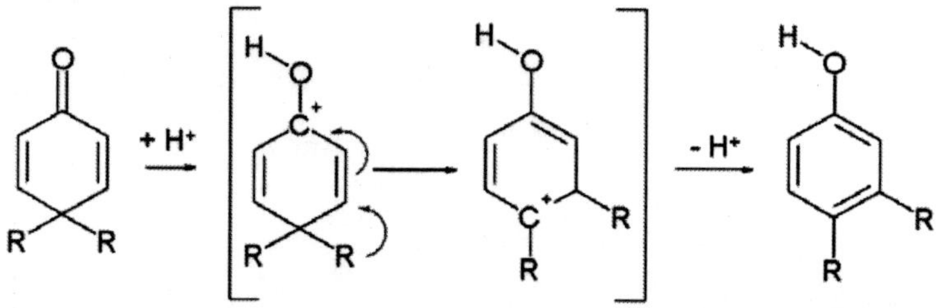

- by the oxidation of aryl silanes—an aromatic variation of the Fleming-Tamao oxidation
- by the addition of benzene and propene in H_3PO_4 to form cumene then O_2 is added with H_2SO_4 to form phenol (Hock Process)
- enzymatic polymerization

Reactions of Phenols

Phenols react in a wide variety of ways.

- Esterfication reactions and ether formation
- Electrophilic aromatic substitutions as the hydroxyl group is activating, for example synthesis of calixarenes
- Reaction of naphtols and hydrazines and sodium bisulfite in the Bucherer carbazole synthesis
- Oxidative cleavage, for instance cleavage of 1,2-dihydroxybenzene to the monomethylester of 2,4 hexadienedioic acid with oxygen, copper chloride in pyridine
- Oxidative de-aromatization to quinones also known as the Teuber reaction. Oxidizing reagents are Fremy's salt and oxone. In reaction depicted below 3,4,5-trimethylphenol reacts with singlet oxygen generated from oxone/sodium carbonate in an acetonitrile/ water mixture to a para-peroxyquinole. This hydroperoxide is reduced to the quinole with sodium thiosulfate.

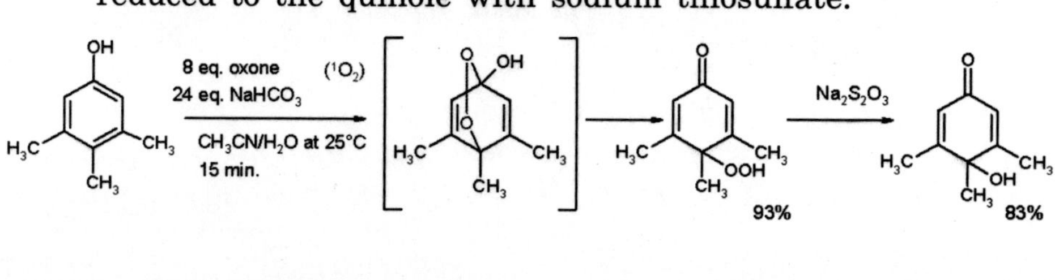

- Phenols are oxidized to hydroquinones in the Elbs persulfate oxidation
- Phenolate anions (deriving from phenols by the loss of H^+) can act as ligands towards metal cations.

Biodegradation

The white rot fungus Phanerochaete chrysosporium can remove up to 80% of phenolic compounds from coking waste water.

Applications

Phenols are important raw materials and additives for industrial purposes in:

- laboratory processes
- chemical industry
- chemical engineering processes
- wood processing
- plastics processing

Tannins are used in the tanning industry.

Some natural phenols can be used as biopesticides. Furanoflavonoids like karanjin or rotenoids are used as acaricide or insecticide.

Enological tannins are important elements in the flavour of wine.

Some phenols are sold as dietary supplements. Phenols have been investigated as drugs. For instance, Crofelemer (USAN, trade name Fulyzaq) is a drug under development for the treatment of diarrhea associated with anti-HIV drugs. Additionally, derivatives have been made of phenolic compound, combretastatin A-4, an anticancer molecule, including nitrogen or halogens atoms to increase the efficacy of the treatment.

Industrial Processing and Analysis

Phenol Extraction

Phenol extraction is a processing technology used to prepare phenols as raw materials, compounds or additives for industrial wood processing and for chemical industries.

Extraction can be performed using different solvents. There is a risk that polyphenol oxidase (PPO) degrades the phenolic content of the sample therefore there is a need to use PPO inhibitors like potassium dithionite ($K_2S_2O_4$) or to perform experiment using liquid nitrogen or to boil the sample for a few seconds (blanching) to inactivate the enzyme. Further fractionation of the extract can be achieved using solid phase extraction columns, and may lead to isolation of individual compounds. The recovery of natural phenols from biomass residue is part of biorefining.

Separation

pK_a of phenolic compounds can be calculated from the retention time in liquid chromatography.

Analytical Methods

Studies on evaluating antioxidant capacity can used electrochemical methods.

Detection can be made by recombinant luminescent bacterial sensors.

Profiling

Phenolic profiling can be achieved with liquid chromatography–mass spectrometry (LC/MS).

Quantification

A method for phenolic content quantification is volumetric titration. An oxidizing agent, permanganate, is used to oxidize known concentrations of a standard solution, producing a standard curve. The content of the unknown phenols is then expressed as equivalents of the appropriate standard. Some methods for quantification of total phenolic content are based on colourimetric measurements. Total phenols (or antioxidant effect) can be measured using the Folin-Ciocalteu reaction. Results are typically expressed as *gallic acid equivalents* (GAE). Ferric chloride ($FeCl_3$) test is also a colourimetric assay.

Lamaison and Carnet have designed a test for the determination of the total flavonoid content of a sample ($AlCI_3$ method). After proper mixing of the sample and the reagent, the mixture is incubated for 10 minutes at ambient temperature and the absorbance of the solution is read at 440 nm. Flavonoid content is expressed in mg/g of quercetin.

Quantitation results produced by the mean of diode array detector-coupled HPLC are generally given as relative rather than absolute values as there is a lack of commercially available standards for every phenolic molecules. The technique can also be coupled with mass spectrometry (for example, HPLC–DAD–ESI/MS) for more precise molecule identification.

Antioxidant Effect Assessment

Other tests measure the antioxidant capacity of a fraction. Some make use of the 2,2'-azino-bis(3-ethylbenzothiazoline-6-sulphonic acid) (ABTS) radical cation, which is reactive towards most antioxidants including phenolics, thiols and vitamin C. During this reaction, the blue ABTS radical cation is converted back to its colourless neutral form. The reaction may be monitored spectrophotometrically. This assay is often referred to as the Trolox equivalent antioxidant capacity (TEAC) assay. The reactivity of the various antioxidants tested are compared to that of Trolox, which is a vitamin E analog.

Other antioxidant capacity assays that use Trolox as a standard include the diphenylpicrylhydrazyl (DPPH), oxygen radical absorbance capacity (ORAC), ferric reducing ability of plasma (FRAP) assays or inhibition of copper-catalyzed *in vitro* human low-density lipoprotein oxidation. A cellular antioxidant activity (CAA) assay also exists. Dichlorofluorescin is a probe that is trapped within cells and is easily oxidized to fluorescent dichlorofluorescein (DCF). The method measures the ability of compounds to prevent the formation of DCF by 2,2'-Azobis(2-amidinopropane) dihydrochloride (ABAP)-generated peroxyl radicals in human hepatocarcinoma HepG2 cells.

The model animal *Galleria mellonella*, the greater waxworm, can be used to test the antioxidant effect of individual molecules using boric acid in food to induce induced an oxidative stress. The content of malondialdehyde, an oxidative stress indicator, and activities of the antioxidant enzymes superoxide dismutase, catalase, glutathione S-transferase and glutathione peroxidase can be monitored. A prophenoloxidase can also be recovered from the insect.

Genetic Analysis

The phenolic biosynthetic and metabolic pathways and enzymes can be studied by mean of transgenesis of genes. The *Arabidopsis* regulatory gene for production of Anthocyanin Pigment 1 (AtPAP1) can be expressed in other plant species.

Natural Occurrences

Phenols are found in the natural world, especially in the plant kingdom.

Occurrences in Prokaryotes

Orobol can be found in *Streptomyces neyagawaensis* (an Actinobacterium). Phenolic compounds can be found in the cyanobacterium *Arthrospira maxima*, used in the dietary supplement, Spirulina. The three cyanobacteria *Microcystis aeruginosa*, *Cylindrospermopsis raciborskii* and *Oscillatoria sp.* are the subject of research into the natural production of butylated hydroxytoluene (BHT), an antioxidant, food additive and industrial chemical.

The proteobacterium *Pseudomonas fluorescens* produces phloroglucinol, phloroglucinol carboxylic acid and diacetylphloroglucinol. Another example of phenolics produced in proteobacteria is 3,5-dihydroxy-4-isopropyl-trans-stilbene, a bacterial stilbenoid produced in *Photorhabdus* bacterial symbionts of *Heterorhabditis* nematodes.

Occurrences in Fungi

Phenolic acids can be found in mushroom basidiomycetes species. For example, protocatechuic acid and pyrocatechol are found in *Agaricus bisporus* as well as other phenylated substances like phenylacetic and phenylpyruvic acids. Other compounds like atromentin and thelephoric acid can also be isolated from fungi in the Agaricomycetes class. Orobol, an isoflavone, can be isolated from *Aspergillus niger*.

Occurrences in Lichen

Gyrophoric acid, a depside, and orcinol are found in lichen.

Occurrence in Algae

The green alga *Botryococcus braunii* is the subject of research into the natural production of butylated hydroxytoluene (BHT), an antioxidant, food additive and industrial chemical.

Phenolic acids such as protocatechuic, p-hydroxybenzoic, 2,3-dihydroxybenzoic, chlorogenic, vanillic, caffeic, *p*-coumaric and salicylic acid, cinnamic acid and hydroxybenzaldehydes such as p-hydroxybenzaldehyde, 3,4-dihydroxybenzaldehyde, vanillin have been isolated from in vitro culture of the freshwater green alga *Spongiochloris spongiosa*.

Phlorotannins, for instance eckol, are found in brown algae. Vidalenolone can be found in the tropical red alga *Vidalia sp*.

Occurrence in Land Plants (embryophytes)

Occurrences in Vascular Plants: Phenolic compounds are mostly found in vascular plants (tracheophytes) i.e. Lycopodiophyta (lycopods), Pteridophyta (ferns and horsetails), Angiosperms (flowering plants or Magnoliophyta) and Gymnosperms (conifers, cycads, Ginkgo and Gnetales).

In ferns, compounds such as kaempferol and its glucoside can be isolated from the methanolic extract of fronds of *Phegopteris connectilis* or kaempferol-3-O-rutinoside, a known bitter-tasting flavonoid glycoside, can be isolated from the rhizomes of *Selliguea feei*. Hypogallic acid, caffeic acid, paeoniflorin and pikuroside can be isolated from the freshwater fern *Salvinia molesta*.

In conifers (Pinophyta), phenolics are stored in polyphenolic parenchyma cells, a tissue abundant in the phloem of all conifers.

The aquatic plant *Myriophyllum spicatum* produces ellagic, gallic and pyrogallic acids and (+)-catechin.

Occurrences in Monocotyledons

Alkylresorcinols can be found in cereals. 2,4-Bis(4-hydroxybenzyl) phenol is a phenolic compound found in the orchids *Gastrodia elata* and *Galeola faberi*.

Occurrences in Non-vascular Plants

Phenolics can also be found in non-vascular land plants (bryophytes). Dihydrostilbenoids and bis(bibenzyls) can be found in liverworts (Marchantiophyta), for instance, the macrocycles cavicularin and riccardin C. Though lignin is absent in mosses (Bryophyta) and hornworts (Anthocerotophyta), some phenolics can be found in those two taxa. For instance, rosmarinic acid and a rosmarinic acid 3'-O-β-D-glucoside can be found in the hornwort *Anthoceros agrestis*.

Occurrences in other Eukaryotes

Occurrences in Insects

The hardening of the protein component of insect cuticle has been shown to be due to the tanning action of an agent produced by oxidation of a phenolic substance forming sclerotin. In the analogous hardening of the cockroach ootheca, the phenolic substance concerned is 3:4-dihydroxybenzoic acid (protocatechuic acid).

Acetosyringone is produced by the male leaffooted bug (*Leptoglossus phyllopus*) and used in its communication system. Guaiacol is produced in the gut of Desert locusts, *Schistocerca gregaria*, by the breakdown of plant material. This process is undertaken by the gut bacterium *Pantoea agglomerans*. Guaiacol is one of the main components of the pheromones that cause locust swarming. Orcinol has been detected in the "toxic glue" of the ant species *Camponotus saundersi*. *Rhynchophorus ferrugineus* (red palm weevil) use 2-methoxy-4-vinylphenol for chemical signaling (pheromones). Other simple and complex phenols can be found in eusocial ants (such as Crematogaster) as components of venom.

Occurrences in Mammals

In female elephants, the two compounds 3-ethyl phenol and 2-ethyl 4,5 dimethylphenol have been detected in urine samples. Temporal glands secretion examination showed the presence of phenol, m-cresol and p-cresol (4-methyl phenol) during musth in male elephants.

p-Cresol and o-cresol are also components of the human sweat. P-cresol is also a major component in pig odor.

4-Ethylphenol, 1,2-dihydroxybenzene, 3-hydroxyacetophenone, 4-methyl-1,2-dihydroxybenzene, 4-methoxyacetophenone, 5-methoxysalicylic acid, salicylaldehyde, and 3-hydroxybenzoic acid are components of castoreum, the exudate from the castor sacs of the mature North American beaver (*Castor canadensis*) and the European beaver (*Castor fibre*), used in perfumery.

Roles

In some cases of natural phenols, they are present in vegetative foliage to discourage herbivory, such as in the case of Western poison oak.

Role in Soils

In soils, it is assumed that larger amounts of phenols are released from decomposing plant litter rather than from throughfall in any natural plant community. Decomposition of dead plant material causes complex organic compounds to be slowly oxidized lignin-like humus or to break down into simpler forms (sugars and amino sugars, aliphatic and phenolic organic acids), which are further transformed into microbial biomass (microbial humus) or are reorganized, and further oxidized, into humic assemblages (fulvic and humic acids), which bind to clay minerals and metal hydroxides. There has been a long debate about the ability of plants to uptake humic substances from their root systems and to metabolize them. There is now a consensus about how humus plays a hormonal role rather than simply a nutritional role in plant physiology.

In the soil, soluble phenols face four different fates. They might be degraded and mineralized as a carbon source by heterotrophic microorganisms; they can be transformed into insoluble and recalcitrant humic substances by polymerization and condensation reactions (with the contribution of soil organisms); they might adsorb to clay minerals or form chelates with aluminium or iron ions; or they might remain in dissolved form, leached by percolating water, and finally leave the ecosystem as part of dissolved organic carbon (DOC).

Leaching is the process by which cations such as iron (Fe) and aluminum (Al), as well as organic matter are removed from the litterfall and transported downward into the soil below. This process is known as podzolization and is particularly intense in boreal and cool temperate forests that are mainly constituted by coniferous pines whose litterfall is rich in phenolic compounds and fulvic acid.

Role in Survival

Phenolic compounds can act as protective agents, inhibitors, natural animal toxicants and pesticides against invading organisms, i.e. herbivores, nematodes, phytophagous insects, and fungal and bacterial pathogens. The scent and pigmentation conferred by other phenolics can attract symbiotic microbes, pollinators and animals that disperse fruits.

Defence against Predators

Volatile phenolic compounds are found in plant resin where they may attract benefactors such as parasitoids or predators of the herbivores that attack the plant.

In the kelp species *Alaria marginata*, phenolics act as chemical defence against herbivores. In tropical *Sargassum* and *Turbinaria* species that are often preferentially consumed by herbivorous fishes and echinoids, there is a relatively low level of phenolics and tannins. Marine allelochemicals generally are present in greater quantity and diversity in tropical than in temperate regions. Marine algal phenolics have been reported as an apparent exception to this biogeographic trend. High phenolic concentrations occur in brown algae species (orders Dictyotales and Fucales) from both temperate and tropical regions, indicating that latitude alone is not a reasonable predictor of plant phenolic concentrations.

Defence against Infection

In *Vitis vinifera* grape, *trans*-resveratrol is a phytoalexin produced against the growth of fungal pathogens such as *Botrytis cinerea* and delta-viniferin is another grapevine phytoalexin produced following fungal infection by *Plasmopara viticola*. Pinosylvin is a pre-infectious stilbenoid toxin (i.e. synthesized prior to infection), contrary to phytoalexins, which are synthesized during infection. It is present in the heartwood of Pinaceae. It is a fungitoxin protecting the wood from fungal infection.

Sakuranetin is a flavanone, a type of flavonoid. It can be found in *Polymnia fruticosa* and rice, where it acts as a phytoalexin against spore germination of *Pyricularia oryzae*. In *Sorghum*, the *SbF3'H2* gene, encoding a flavonoid 3'-hydroxylase, seems to be expressed in pathogen-specific 3-deoxyanthocyanidin phytoalexins synthesis, for example in *Sorghum-Colletotrichum* interactions.

6-Methoxymellein is a dihydroisocoumarin and a phytoalexin induced in carrot slices by UV-C, that allows resistance to *Botrytis cinerea* and other microorganisms.

Danielone is a phytoalexin found in the papaya fruit. This compound showed high antifungal activity against *Colletotrichum gloesporioides*, a pathogenic fungus of papaya. Stilbenes are produced in *Eucalyptus sideroxylon* in case of pathogens attacks. Such compounds can be implied in the hypersensitive response of plants. High levels of phenolics in some woods can explain their natural preservation against rot.

In plants, VirA is a protein histidine kinase which senses certain sugars and phenolic compounds. These compounds are typically found from wounded plants, and as a result VirA is used by *Agrobacterium tumefaciens* to locate potential host organisms for infection.

Role in Allelopathic Interactions

Natural phenols can be involved in allelopathic interactions, for example in soil or in water. Juglone is an example of such a molecule inhibiting the growth of other plant species around walnut trees. The green alga *Myriophyllum spicatum* produces ellagic, gallic and pyrogallic acids and (+)-catechin, allelopathic phenolic compounds inhibiting the growth of blue-green alga *Microcystis aeruginosa*.

Phenolics, and in particular flavonoids and isoflavonoids, may be involved in endomycorrhizae formation.

Acetosyringone has been best known for its involvement in plant-pathogen recognition, especially its role as a signal attracting and transforming unique, oncogenic bacteria in genus *Agrobacterium*. The virA gene on the Ti plasmid in the genome of *Agrobacterium tumefaciens* and *Agrobacterium rhizogenes* is used by these soil bacteria to infect plants, via its encoding for a receptor for acetosyringone and other phenolic phytochemicals exuded by plant wounds. This compound also allows higher transformation efficiency in plants, in A. tumefaciens mediated transformation procedures, and so is of importance in plant biotechnology.

Content in Human Food

Notable sources of natural phenols in human nutrition include berries, tea, beer, olive oil, chocolate or cocoa, coffee, pomegranates, popcorn, yerba maté, fruits and fruit based drinks (including cider, wine and vinegar) and vegetables. Herbs and spices, nuts (walnuts, peanut) and algae are also potentially significant for supplying certain natural phenols. Natural phenols can also be found in fatty matrices like olive oil. Cloudy olive oil has the higher levels of phenols, or polar phenols that form a complex phenol-protein complex.

Phenolic compounds, when used in beverages, such as prune juice, have been shown to be helpful in the colour and sensory components, such as alleviating bitterness. Some advocates for organic farming claim that organically grown potatoes, oranges, and leafy vegetables have more phenolic compounds and these may provide antioxidant protection against heart disease and cancer. However evidence on substantial differences between organic food and conventional food is insufficient to make claims that organic food is safer or more healthy than conventional food.

Dissolved Organic Carbon

Dissolved organic carbon (DOC) is a broad classification for organic molecules of varied origin and composition within aquatic systems. The "dissolved" fraction of organic carbon is an operational classification. Many researchers use the term "dissolved" for compounds below 0.45 micrometres, but 0.22 micrometres is also common, saving colloidal for higher concentrations. A practical definition of dissolved typically used in marine chemistry is all substances that pass through a GF/F filter. The recommended measure technique is the HTCO technique after filtration on precombusted glass fibre filters, typically GF/F filters.

DOC in marine and freshwater systems is one of the greatest cycled reservoirs of organic matter on Earth. The source of DOC depends on the body of water. In general, organic carbon compounds are a result of decomposition processes from dead organic matter such as plants. When water contacts highly organic soils, these components can drain into rivers and lakes as DOC.

DOC is also extremely important in the transport of metals in aquatic systems. Metals form extremely strong complexes with DOC, enhancing metal solubility while also reducing metal bioavailability.

Significance

DOC is a food supplement, supporting growth of microorganisms and plays an important role in the global carbon cycle through the microbial loop. Moreover it is an indicator of organic loadings in streams, as well as supporting terrestrial processing (e.g., within soil, forests, and wetlands) of organic matter. Dissolved organic carbon has a high proportion of biodegradable dissolved organic carbon (BDOC) in first order streams compared to higher order streams. In the absence of extensive wetlands, bogs, or swamps, baseflow concentrations of DOC in undisturbed watersheds generally range from approximately 1 to 20 mg/L carbon. Carbon concentrations considerably vary across ecosystems. For example, the Everglades may be near the top of the range and the middle of oceans may be near the bottom. Occasionally, high concentrations of organic carbon indicate anthropogenic influences, but most DOC originates naturally.

The *BDOC fraction* consists of organic molecules that heterotrophic bacteria can use as a source of energy and carbon. Some subset of DOC constitutes the precursors of disinfection byproducts for drinking water. BDOC can contribute to undesirable biological regrowth within water distribution systems.

Phenolic Compounds in Plants

Like all the groups reviewed until now, this one does also consist of a large number of molecules of heterogeneous structure. Their common feature is the presence of at least one hydroxyl-substituted aromatic ring system.

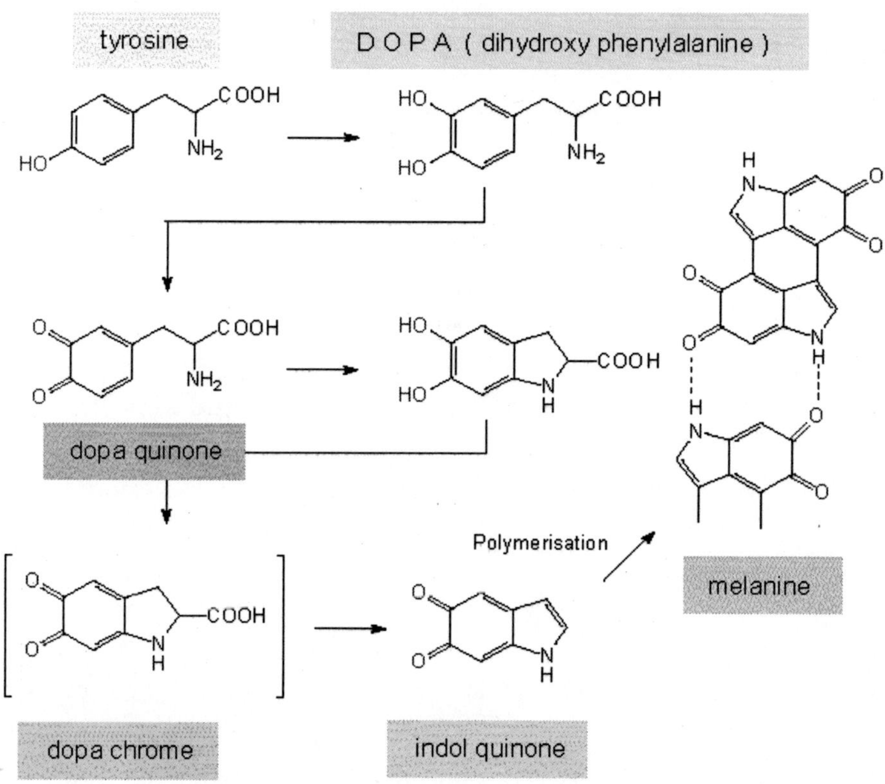

Most phenolic compounds belong to the flavonoids. Lignin, the primary substance of wood, is the most common member of this group. The following table outlines the most important groups of plant phenolic compounds.

Table: *The Most Important Classes of Phenolic Compounds in Plants*

class	*number of basic skeleton*	*C-atoms*
6	C_6	simple phenols, benzoquinones
7	$C_6 - C_1$	phenolic acids
8	$C_6 - C_2$	acetophenone, phenylacetic acid
9	$C_6 - C_3$	hydroxycinnamic acid, polypropene,coumarin, isocoumarin
10	$C_6 - C_4$	naphtoquinone
13	$C_6 - C_1 - C_6$	xanthone
14	$C_6 - C_2 - C_6$	stilbene, anthrachinone
15	$C_6 - C_3 - C_6$	flavonoids, isoflavonoids
18	$(C_6 - C_3)_2$	lignans, neolignans
30	$(C_6 - C_3 - C_6)_2$	biflavonoids
n	$(C_6 - C_3)_n(C_6)_n(C_6 - C_3 - C_6)_n$	lignins catecholmelanine (condensed tannins)

The starting product of the biosynthesis of most phenolic compounds is shikimate. Phenols are acidic due to the dissociability of their -OH group. They are rather reactive compounds and as long as no steric inhibition due to additional side chains occurs, they form hydrogen bonds. Consequently, many flavonoids have intramolecular bonds. Another important feature is their ability to form chelate complexes with metals. Also, they are easily oxidized and, if so, form polymers (dark aggregates). The darkening of cut or dying plant parts is caused by this reaction. They have usually an inhibiting effect on plant growth. Among the phenylpropanol derivatives of lower molecular weight are a number of scents like the coumarins, cinnamic acid, sinapinic acid, the coniferyl alcohols and others. These substances and their derivatives are at the same time intermediates of the biosynthesis of lignin.

Table: *The Most Important Classes of Flavonoids and their Biological Significance*

class	*number of known members*	*biological significance (so far as known)*
anthocyanin(s)	250	red and blue pigments
chalcons	60	yellow pigments
aurones	20	yellow pigments
flavones	350	cream-coloured pigments of flowers
flavonols	350	feeding repellents (?) in leaves
dihydrochalcons	10	some taste bitter
proanthocyanidins	50	astringent substances
catechins	40	some have propertieslike those of tannins
biflavonoids ?	65	?
isoflavonoids	15	oestrogen effect, toxic for fungi

The basic structure of flavonoids is derived from the C_{15} body of flavone. They differ from other phenolic substances in the degree of oxidation of their central pyran ring. And, very fundamentally, also in their biological properties. While some classes (the flavonones, for example) are colourless, the members of other classes (the anthocyanes, for example) are always coloured and known as pigments of flowers or other plant parts. Anthocyanes are normally red or yellow, their colour is pH-dependent. Blue pigments are achieved by chelate formation with certain metal ions.

The variability of the flavonoids is largely based on the hydroxylation and/ or methylation pattern of the three ring systems. A correlation between two flavonoids points often to a relationship between the producing plant species. They have therefore proven to be suitable traits for the study of the phylogenetic relations between higher plants. The quinones are another group of phenolic compounds. We have already met some of its members that function as co-factors. Accordingly, they do not actually belong to the secondary plant products but have to be counted among those of the basic metabolism. As has been mentioned before, phenolic compounds occur usually not unbound within plant tissues. They are mostly coupled to other molecules, often to glucosyl residues, but to sulphate- or acetyl-residues, too. One of the reasons may be that they are toxic when in a free state and are detoxified, at least partially, if coupled. Many low molecular weight compounds, for example thymol, are used in medicine as antiseptics due to their toxicity. Different types of bonds between flavonoids (for example anthocyanes) and a glycosyl residue lead to different derivatives that increase the range of flower colours (and colour shades). The glycosylation of flavonoids has an additional, ecologically not less important function. It has been brought into connection with pest protection and protection against other animals.

5

Terpenoids

Plant allelochemicals released into the soil can significantly impact the performance of associated plant species thereby affecting their competitive ability. Soil microbes can potentially affect the interaction between plant and plant chemicals by degrading the allelochemicals. However, most often plant-plant chemical interactions are studied using filter paper bioassays examining the pair-wise interaction between a plant and a plant chemical, not taking into account the potential role of soil microorganisms.

The terpenoids, sometimes called isoprenoids, are a large and diverse class of naturally occurring organic chemicals similar to terpenes, derived from five-carbon isoprene units assembled and modified in thousands of ways. Most are multicyclic structures that differ from one another not only in functional groups but also in their basic carbon skeletons. These lipids can be found in all classes of living things, and are the largest group of natural products.

Plant terpenoids are used extensively for their aromatic qualities. They play a role in traditional herbal remedies and are under investigation for antibacterial, antineoplastic, and other pharmaceutical functions. Terpenoids contribute to the scent of eucalyptus, the flavors of cinnamon, cloves, and ginger, the yellow colour in sunflowers, and the red colour in tomatoes. Well-known terpenoids include citral, menthol, camphor, salvinorin A in the plant *Salvia divinorum*, the cannabinoids found in cannabis, ginkgolide and bilobalide found in *Ginkgo biloba*, and the curcuminoids found in turmeric and mustard seed.

The steroids and sterols in animals are biologically produced from terpenoid precursors. Sometimes terpenoids are added to proteins, e.g., to enhance their attachment to the cell membrane; this is known as isoprenylation.

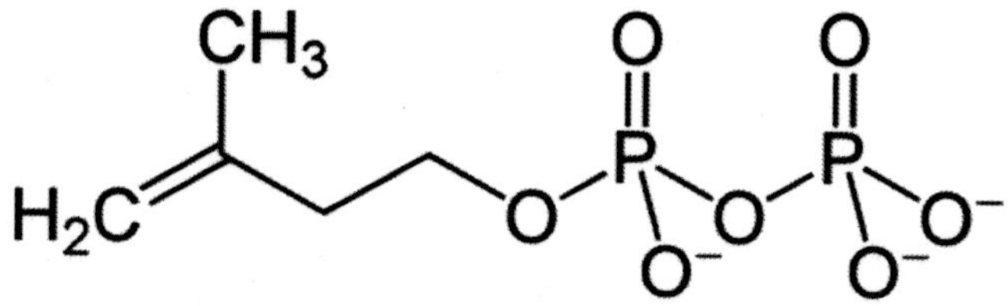

Figure: *Chemical structure of the terpenoid isopentenyl pyrophosphate*

Structure and Classification

Terpenes are hydrocarbons resulting from the combination of several isoprene units. Terpenoids can be thought of as modified terpenes, wherein methyl groups have been moved or removed, or oxygen atoms added. (Some authors use the term "terpene" more broadly, to include the terpenoids.) Just like terpenes, the terpenoids can be classified according to the number of isoprene units used:

- Hemiterpenoids, 1 isoprene unit (5 carbons)
- Monoterpenoids, 2 isoprene units (10C)
- Sesquiterpenoids, 3 isoprene units (15C)
- Diterpenoids, 4 isoprene units (20C) (e.g. ginkgolides)
- Sesterterpenoids, 5 isoprene units (25C)
- Triterpenoids, 6 isoprene units (30C) (e.g. sterols)
- Tetraterpenoids, 8 isoprene units (40C) (e.g. carotenoids)
- Polyterpenoid with a larger number of isoprene units

Terpenoids can also be classified according to the number of cyclic structures they contain. In the year 1995, Julius Anitnec found out that Terpenoids can be tested for using the Salkowski test.

Meroterpenes are any compound, including many natural products, having a partial terpenoid structure

Biosynthesis

There are two metabolic pathways of creating terpenoids:

Mevalonic Acid Pathway: Many organisms manufacture terpenoids through the HMG-CoA reductase pathway, the pathway that also produces cholesterol. The reactions take place in the cytosol. The pathway was discovered in the 1950s.

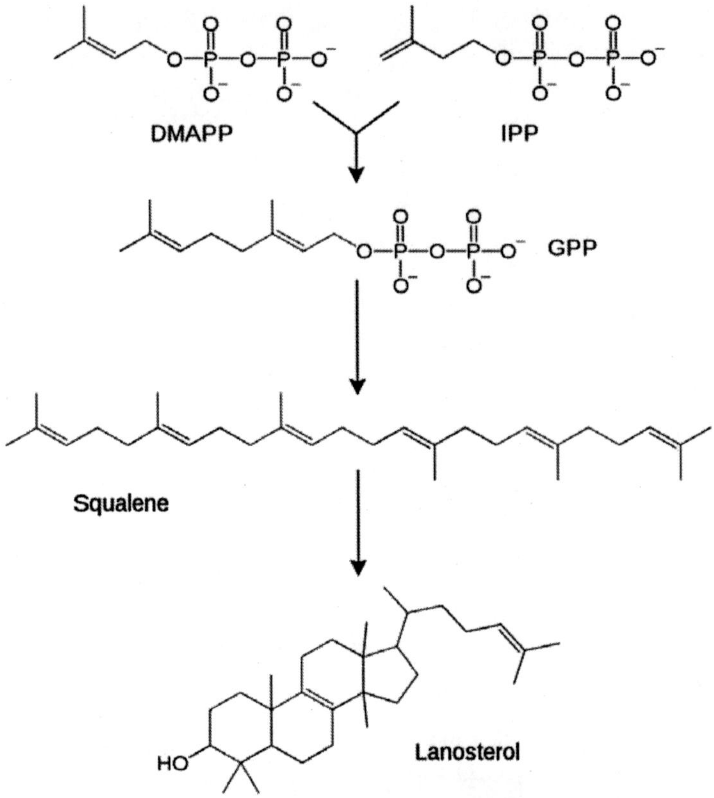

Figure: *Simplified version of the steroid synthesis pathway with the terpenoid intermediates isopentenyl pyrophosphate (IPP), dimethylallyl pyrophosphate (DMAPP), geranyl pyrophosphate (GPP), and squalene shown. Some intermediates are omitted for clarity.*

MEP/DOXP Pathway

The 2-*C*-methyl-D-erythritol 4-phosphate/1-deoxy-D-xylulose 5-phosphate pathway (MEP/DOXP pathway), also known as [non-mevalonate pathway] or mevalonic acid-independent pathway, takes place in the plastids of plants and apicomplexan protozoa, as well as in many bacteria. It was discovered in the late 1980s.

Pyruvate and glyceraldehyde 3-phosphate are converted by DOXP synthase (Dxs) to 1-deoxy-D-xylulose 5-phosphate, and by DOXP reductase (Dxr, IspC) to 2-*C*-methyl-D-erythritol 4-phosphate (MEP). The subsequent three reaction steps catalyzed by 4-diphosphocytidyl-2-*C*-methyl-D-erythritol synthase (YgbP, IspD), 4-diphosphocytidyl-2-*C*-methyl-D-erythritol kinase (YchB, IspE), and 2-*C*-methyl-D-erythritol 2,4-cyclodiphosphate synthase (YgbB, IspF) mediate the formation of 2-*C*-methyl-D-erythritol 2,4-cyclopyrophosphate (MEcPP). Finally, MEcPP is converted to (*E*)-4-hydroxy-3-methyl-but-2-enyl pyrophosphate

(HMB-PP) by HMB-PP synthase (GcpE, IspG), and HMB-PP is converted to isopentenyl pyrophosphate (IPP) and dimethylallyl pyrophosphate (DMAPP) by HMB-PP reductase (LytB, IspH).

IPP and DMAPP are the end-products in either pathway, and are the precursors of isoprene, monoterpenoids (10-carbon), diterpenoids (20-carbon), carotenoids (40-carbon), chlorophylls, and plastoquinone-9 (45-carbon). Synthesis of all higher terpenoids proceeds via formation of geranyl pyrophosphate (GPP), farnesyl pyrophosphate (FPP), and geranylgeranyl pyrophosphate (GGPP).

Although both pathways, MVA and MEP, are mutually exclusive in most organisms, interactions between them have been reported in plants and few bacteria species.

Organism	***Pathways***
Bacteria	MVA or MEP
Archaea	MVA
Green Algae	MEP
Plants	MVA and MEP
Animals	MVA
Fungi	MVA

A study was conducted to understand the effects of certain phenolics, terpenoids, and their equimolar mixture through agar gel and soil growth bioassays and their recovery from soils. The eight compounds selected for this study were *p*-hydroxybenzoic acid, ferulic acid, umbelliferone, catechin, emodin, 1,8-cineole, carvone, and betulin. Lettuce (*Lactuca sativa* L.) was used as test species for agar gel and soil growth bioassays. Root and shoot growth of lettuce was inhibited for all the above except emodin and catechin. However, in soils treated with different phenolics and terpenoids, only root growth of lettuce was inhibited, whereas shoot growth was promoted.

Terpene

Terpenes are widespread in nature, mainly in plants as constituents of essential oils. Many terpenes are hydrocarbons, but oxygen-containing compounds such as alcohols, aldehydes or ketones (*terpenoids*) are also found.

Terpenes are organic compounds, produced by a variety of plants, particularly conifers, though also by some insects such as termites or swallowtail butterflies, which emit terpenes from their osmetreia. They are often strong-smelling, and thus may protect the plants that produce them by deterring parasites. Many terpenes are aromatic hydrocarbons and thus may have had a protective function. The difference between terpenes and terpenoids is that terpenes are hydrocarbons, whereas terpenoids contain additional functional groups.

They are the major components of resin, and of turpentine produced from resin. The name "terpene" is derived from the word "turpentine". In addition to their roles as end-products in many organisms, terpenes are major biosynthetic building blocks within nearly every living creature. Steroids, for example, are derivatives of the triterpene squalene. When terpenes are modified chemically, such as by oxidation or rearrangement of the carbon skeleton, the resulting compounds are generally referred to as terpenoids. Some authors will use the term terpene to include all terpenoids. Terpenoids are also known as isoprenoids.

Terpenes and terpenoids are the primary constituents of the essential oils of many types of plants and flowers. Essential oils are used widely as natural flavour additives for food, as fragrances in perfumery, and in traditional and alternative medicines such as aromatherapy. Synthetic variations and derivatives of natural terpenes and terpenoids also greatly expand the variety of aromas used in perfumery and flavors used in food additives. Vitamin A is a terpene. Terpenes are released by trees more actively in warmer weather, acting as a natural form of cloud seeding. The clouds reflect sunlight, allowing the forest to regulate its temperature.

The aroma and flavour of hops, highly desirable in some beers, comes from terpenes. Of the terpenes in hops myrcene, β-pinene, β-caryophyllene, and α-humulene are found in the largest quantities.

Structure and Biosynthesis

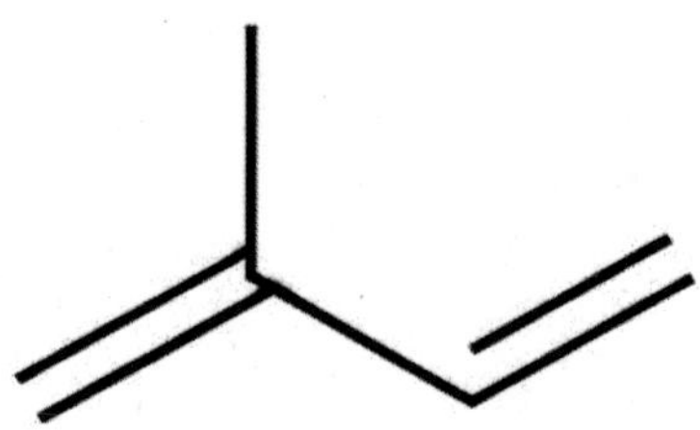

***Figure:** Isoprene*

Terpenes are derived biosynthetically from units of isoprene, which has the molecular formula C_5H_8. The basic molecular formulae of terpenes are multiples of that, $(C_5H_8)_n$ where n is the number of linked isoprene units. This is called the isoprene rule or the *C5 rule*. The isoprene units may be linked together "head to tail" to form linear chains or they may be arranged to form rings. One can consider the isoprene unit as one of nature's common building blocks.

Isoprene itself does not undergo the building process, but rather activated forms, isopentenyl pyrophosphate (IPP or also isopentenyl diphosphate) and dimethylallyl pyrophosphate (DMAPP or also dimethylallyl diphosphate), are the components in the biosynthetic pathway. IPP is formed from acetyl-CoA via the intermediacy of mevalonic acid in the HMG-CoA reductase pathway. An alternative, totally unrelated biosynthesis pathway of IPP is known in some bacterial groups and the plastids of plants, the so-called MEP(2-Methyl-D-erythritol-4-phosphate)-pathway, which is initiated from C5-sugars. In both pathways, IPP is isomerized to DMAPP by the enzyme isopentenyl pyrophosphate isomerase.

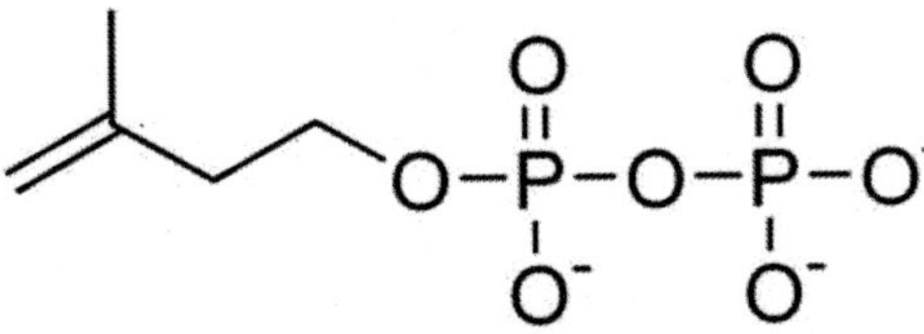

***Figure:** Isopentenyl pyrophosphate*

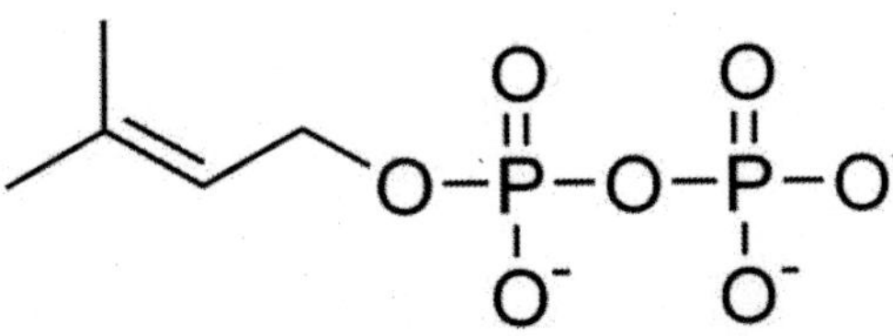

***Figure:** Dimethylallyl pyrophosphate*

As chains of isoprene units are built up, the resulting terpenes are classified sequentially by size as hemiterpenes, monoterpenes, sesquiterpenes, diterpenes, sesterterpenes, triterpenes, and tetraterpenes. Essentially, they are all synthesised by terpene synthase.

Types

***Figure:** Second or third instar caterpillars of* Papilio glaucus *emit terpenes from their osmeterium.*

Terpenes may be classified by the number of isoprene units in the molecule; a prefix in the name indicates the number of terpene units needed to assemble the molecule.

- Hemiterpenes consist of *a single isoprene* unit. Isoprene itself is considered the only hemiterpene, but oxygen-containing derivatives such as prenol and isovaleric acid are hemiterpenoids.
- Monoterpenes consist of *two isoprene* units and have the molecular formula $C_{10}H_{16}$. Examples of monoterpenes are: geraniol, limonene and terpineol.
- Sesquiterpenes consist of *three isoprene* units and have the molecular formula $C_{15}H_{24}$. Examples of sesquiterpenes are: humulene, farnesenes, farnesol. (The *sesqui-* prefix means one and a half.)
- Diterpenes are composed of *four isoprene* units and have the molecular formula $C_{20}H_{32}$. They derive from geranylgeranyl pyrophosphate. Examples of diterpenes are cafestol, kahweol, cembrene and taxadiene (precursor of taxol). Diterpenes also form the basis for biologically important compounds such as retinol, retinal, and phytol. They are known to be antimicrobial and antiinflammatory.
- Sesterterpenes, terpenes having 25 carbons and *five isoprene* units, are rare relative to the other sizes. (The *sester-* prefix means half to three, i.e. two and a half.) An example of a sesterterpene is geranylfarnesol.
- Triterpenes consist of *six isoprene* units and have the molecular formula $C_{30}H_{48}$. The linear triterpene squalene, the major constituent of shark liver oil, is derived from the reductive coupling of two molecules of farnesyl pyrophosphate. Squalene is then processed biosynthetically to generate either lanosterol or cycloartenol, the structural precursors to all the steroids.
- Sesquarterpenes are composed of *seven isoprene* units and have the molecular formula $C_{35}H_{56}$. Sesquarterpenes are typically microbial in their origin. Examples of sesquarterpenes are ferrugicadiol and tetraprenylcurcumene.
- Tetraterpenes contain *eight isoprene* units and have the molecular formula $C_{40}H_{64}$. Biologically important tetraterpenes include the acyclic lycopene, the monocyclic gamma-carotene, and the bicyclic alpha- and beta-carotenes.
- Polyterpenes consist of long chains of *many isoprene* units. Natural rubber consists of polyisoprene in which the double

bonds are cis. Some plants produce a polyisoprene with trans double bonds, known as gutta-percha.

- Norisoprenoids, such as the C_{13}-norisoprenoids 3-oxo-α-ionol present in Muscat of Alexandria leaves and 7,8-dihydroionone derivatives, such as megastigmane-3,9-diol and 3-oxo-7,8-dihydro-α-ionol found in Shiraz leaves (both grapes in the species *Vitis vinifera*) or wine (responsible for some of the spice notes in Chardonnay), can be produced by fungal peroxydases or glycosidases.

Other Uses

Research into terpenes has found that many of them possess qualities that make them ideal active ingredients as part of natural agricultural pesticides. Terpin hydrate is a derivative of turpentine. An expectorant and humectant, it is commonly used in the treatment of acute or chronic bronchitis and related conditions. Terpenes are used by termites of the Nasutitermitinae family to attack enemy insects, through the use of a specialized mechanism called a fontanellar gun.

The band Carbon Based Lifeforms has a song called Terpenes on their album Twentythree.

Alkaloids

Alkaloids are a group of naturally occurring chemical compounds that contain mostly basic nitrogen atoms. This group also includes some related compounds with neutral and even weakly acidic properties. Some synthetic compounds of similar structure are also attributed to alkaloids. In addition to carbon, hydrogen and nitrogen, alkaloids may also contain oxygen, sulfur and more rarely other elements such as chlorine, bromine, and phosphorus.

Alkaloids are produced by a large variety of organisms, including bacteria, fungi, plants, and animals, and are part of the group of natural products (also called secondary metabolites). Many alkaloids can be purified from crude extracts by acid-base extraction. Many alkaloids are toxic to other organisms. They often have pharmacological effects and are used as medications, as recreational drugs, or in entheogenic rituals. Examples are the local anesthetic and stimulant cocaine, the psychedelic psilocin, the stimulant caffeine, nicotine, the analgesic morphine, the antibacterial berberine, the anticancer compound vincristine, the antihypertension agent reserpine, the cholinomimetic galantamine, the anticholinergic agent atropine, the vasodilator vincamine, the antiarrhythmia compound quinidine, the antiasthma therapeutic ephedrine, and the antimalarial drug quinine. Although alkaloids act on a diversity of metabolic systems in humans and other animals, they almost uniformly invoke a bitter taste.

The boundary between alkaloids and other nitrogen-containing natural compounds is not clear-cut. Compounds like amino acid peptides, proteins, nucleotides, nucleic acid, amines, and antibiotics are usually not called alkaloids. Natural compounds containing nitrogen in the exocyclic position (mescaline, serotonin, dopamine, etc.) are usually attributed to amines rather than alkaloids. Some authors, however, consider alkaloids a special case of amines.

History

Alkaloid-containing plants have been used by humans since ancient times for therapeutic and recreational purposes. For example, medicinal plants have been known in the Mesopotamia at least around 2000 BC. The *Odyssey* of Homer referred to a gift given to Helen by the Egyptian queen, a drug bringing oblivion. It is believed that the gift was an opium-containing drug. A Chinese book on houseplants written in 1st–3rd centuries BC mentioned a medical use of Ephedra and opium poppies. Also, coca leaves have been used by South American Indians since ancient times.

Extracts from plants containing toxic alkaloids, such as aconitine and tubocurarine, were used since antiquity for poisoning arrows.

Studies of alkaloids began in the 19th century. In 1804, the German chemist Friedrich Sertürner isolated from opium a "soporific principle" (Latin: *principium somniferum*), which he called "morphium" in honour of Morpheus, the Greek god of dreams; in German and some other Central-European languages, this is still the name of the drug. The term "morphine", used in English and French, was given by the French physicist Joseph Louis Gay-Lussac.

A significant contribution to the chemistry of alkaloids in the early years of its development was made by the French researchers Pierre Joseph Pelletier and Joseph Bienaimé Caventou, who discovered quinine (1820) and strychnine (1818). Several other alkaloids were discovered around that time, including xanthine (1817), atropine (1819), caffeine (1820), coniine (1827), nicotine (1828), colchicine (1833), sparteine (1851), and cocaine (1860).

The first complete synthesis of an alkaloid was achieved in 1886 by the German chemist Albert Ladenburg. He produced coniine by reacting 2-methylpyridine with acetaldehyde and reducing the resulting 2-propenyl pyridine with sodium. The development of the chemistry of alkaloids was accelerated by the emergence of spectroscopic and chromatographic methods in the 20th century, so that by 2008 more than 12,000 alkaloids had been identified.

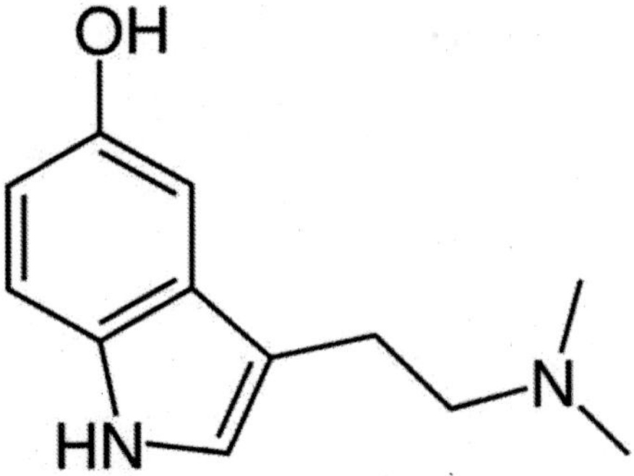

***Figure:** Bufotenin, an alkaloid from some toads, contains an indole core and is produced in living organisms from the amino acid tryptophan.*

Properties

Most alkaloids contain oxygen in their molecular structure; those compounds are usually colourless crystals at ambient conditions. Oxygen-free alkaloids, such as nicotine or coniine, are typically volatile, colourless, oily liquids. Some alkaloids are coloured, like berberine (yellow) and sanguinarine (orange).

Most alkaloids are weak bases, but some, such as theobromine and theophylline, are amphoteric. Many alkaloids dissolve poorly in water but readily dissolve in organic solvents, such as diethyl ether, chloroform or 1,2-dichloroethane. Caffeine, cocaine, codeine and nicotine are water soluble (with a solubility of e–1g/L), whereas others, including morphine and yohimbine are highly water soluble (0.1–1 g/L). Alkaloids and acids form salts of various strengths. These salts are usually soluble in water and alcohol and poorly soluble in most organic solvents. Exceptions include scopolamine hydrobromide, which is soluble in organic solvents, and the water-soluble quinine sulfate.

Most alkaloids have a bitter taste or are poisonous when ingested. Alkaloid production in plants appeared to have evolved in response to feeding by herbivorous animals; however, some animals have evolved the ability to detoxify alkaloids. Some alkaloids can produce developmental defects in the offspring of animals that consume but cannot detoxify the alkaloids. One example is the alkaloid cyclopamine, produced in the leaves of corn lily. During the 1950s, up to 25% of lambs born by sheep that had grazed on corn lily had serious facial deformations. These ranged from deformed jaws to cyclopia. After decades of research, in the 1980s, the compound responsible for these deformities was identified as the alkaloid 11-deoxyjervine, later renamed to cyclopamine.

Distribution in Nature

Alkaloids are generated by various living organisms, especially by higher plants – about 10 to 25% of those contain alkaloids. Therefore, in the past the term "alkaloid" was associated with plants.

The alkaloids content in plants is usually within a few percent and is inhomogeneous over the plant tissues. Depending on the type of plants, the maximum concentration is observed in the leaves (black henbane), fruits or seeds (Strychnine tree), root (Rauwolfia serpentina) or bark (cinchona). Furthermore, different tissues of the same plants may contain different alkaloids.

Beside plants, alkaloids are found in certain types of fungi, such as psilocybin in the fungus of the genus Psilocybe, and in animals, such as bufotenin in the skin of some toads. Many marine organisms also contain alkaloids. Some amines, such as adrenaline and serotonin, which play an important role in higher animals, are similar to alkaloids in their structure and biosynthesis and are sometimes called alkaloids.

Extraction

***Figure:** Crystals of piperine extracted from black pepper.*

Because of the structural diversity of alkaloids, there is no single method of their extraction from natural raw materials. Most methods exploit the property of most alkaloids to be soluble in organic solvents but not in water, and the opposite tendency of their salts.

Most plants contain several alkaloids. Their mixture is extracted first and then individual alkaloids are separated. Plants are thoroughly ground before extraction. Most alkaloids are present in the raw plants in the form of salts of organic acids. The extracted alkaloids may remain salts or change into bases. Base extraction is achieved by processing the raw material with alkaline solutions and extracting the alkaloid bases with organic solvents, such as 1,2-dichloroethane, chloroform, diethyl ether or benzene. Then, the impurities are dissolved by weak acids; this converts alkaloid bases into salts that are washed away with water. If necessary, an aqueous solution of alkaloid salts is again made alkaline and treated with an organic solvent. The process is repeated until the desired purity is achieved.

In the acidic extraction, the raw plant material is processed by a weak acidic solution (e.g., acetic acid in water, ethanol, or methanol). A base is then added to convert alkaloids to basic forms that are extracted with organic solvent (if the extraction was performed with alcohol, it is removed first, and the remainder is dissolved in water). The solution is purified as described above. Alkaloids are separated from their mixture using their different solubility in certain solvents and different reactivity with certain reagents or by distillation.

Biosynthesis

Biological precursors of most alkaloids are amino acids, such as ornithine, lysine, phenylalanine, tyrosine, tryptophan, histidine, aspartic acid, and anthranilic acid. Nicotinic acid can be synthesized from tryptophan or aspartic acid. Ways of alkaloid biosynthesis are too numerous and cannot be easily classified. However, there are a few typical reactions involved in the biosynthesis of various classes of alkaloids, including synthesis of Schiff bases and Mannich reaction.

Synthesis of Schiff Bases

Schiff bases can be obtained by reacting amines with ketones or aldehydes. These reactions are a common method of producing C=N bonds.

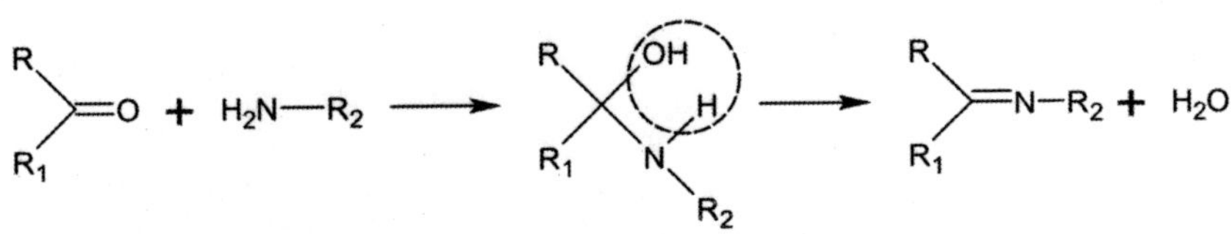

In the biosynthesis of alkaloids, such reactions may take place within a molecule, such as in the synthesis of piperidine:

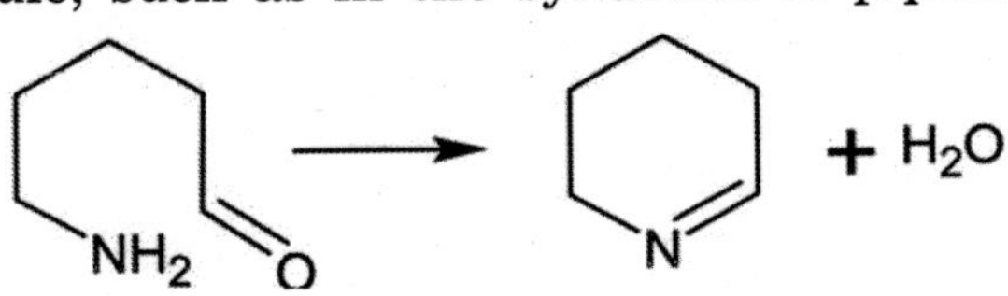

Mannich Reaction

An integral component of the Mannich reaction, in addition to an amine and a carbonyl compound, is a carbanion, which plays the role of the nucleophile in the nucleophilic addition to the ion formed by the reaction of the amine and the carbonyl.

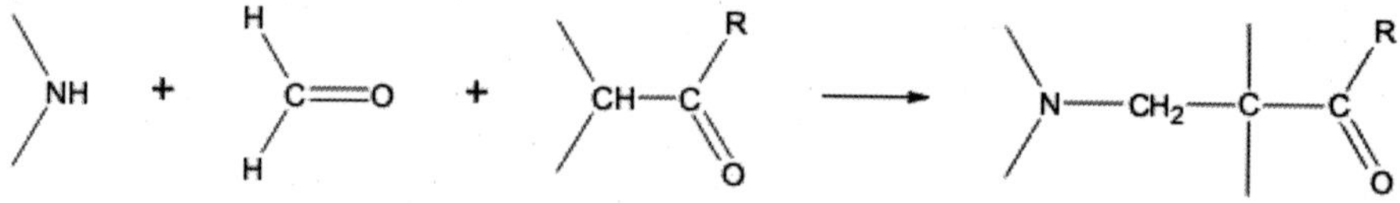

The Mannich reaction can proceed both intermolecularly and intramolecularly:

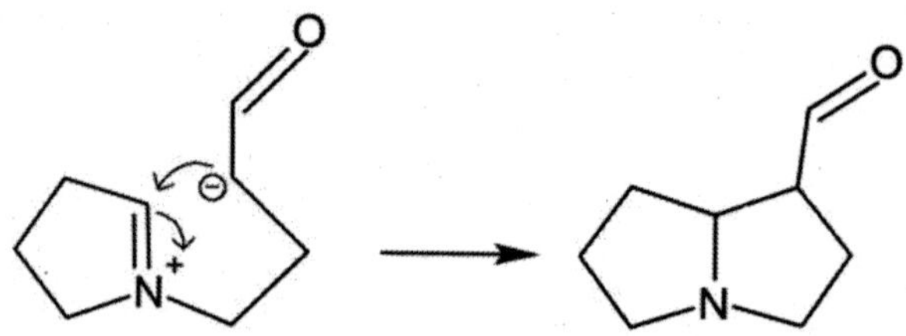

Dimer Alkaloids

In addition to the described above monomeric alkaloids, there are also dimeric, and even trimeric and tetrameric alkaloids formed upon condensation of two, three, and four monomeric alkaloids. Dimeric alkaloids are usually formed from monomers of the same type through the following mechanisms:

- Mannich reaction, resulting in, e.g., voacamine
- Michael reaction (villalstonine)
- Condensation of aldehydes with amines (toxiferine)
- Oxidative addition of phenols (dauricine, tubocurarine)
- Lactonization (carpaine).

The Biological Role

The role of alkaloids for living organisms that produce them is still unclear. It was initially assumed that the alkaloids are the final products of nitrogen metabolism in plants, as urea in mammals. It was later shown that alkaloid concentrations varies over time, and this hypothesis was refuted.

Most of the known functions of alkaloids are related to protection. For example, aporphine alkaloid liriodenine produced by the tulip tree protects it from parasitic mushrooms. In addition, presence of alkaloids in the plant prevents insects and chordate animals from eating it. However, some animals adapted to alkaloids and even use them in their own metabolism. Such alkaloid-related substances as serotonin, dopamine and histamine are important neurotransmitters in animals. Alkaloids are also known to regulate plant growth. Another example of an organism that uses alkaloids for protection is the *Utetheisa ornatrix*, more commonly known as the Ornate Moth. Pyrrolizidine alkaloids render these larvae and adult moths unpalatable to many of their natural enemies like coccinelid beetles, green lacewings, insectivorous hemiptera and insectivorous bats.

Applications

In Medicine: Medical use of alkaloid-containing plants has a long history, and, thus, when the first alkaloids were isolated in the 19th century, they immediately found application in clinical practice. Many alkaloids are still used in medicine, usually in the form of salts, including the following:

Alkaloid	*Action*
Ajmaline	antiarrhythmic
Atropine, scopolamine, hyoscyamine	anticholinergic
Caffeine	Stimulant, Adenosine receptor antagonist
Codeine	cough medicine, analgesic
Colchicine	remedy for gout
Emetine	antiprotozoal agent
Ergot alkaloids	sympathomimetic, vasodilator, antihypertensive
Morphine	analgesic
Nicotine	Stimulant, Nicotinic acetylcholine receptor agonist
Physostigmine	inhibitor of acetylcholinesterase
Quinidine	antiarrhythmic
Quinine	antipyretics, antimalarial
Reserpine	antihypertensive
Tubocurarine	Muscle relaxant
Vinblastine, vincristine	antitumor
Vincamine	vasodilating, antihypertensive
Yohimbine	Stimulant, Aphrodisiac

Many synthetic and semisynthetic drugs are structural modifications of the alkaloids, which were designed to enhance or change the primary effect of the drug and reduce unwanted side-effects. For example, naloxone, an opioid receptor antagonist, is a derivative of thebaine that is present in opium.

In Agriculture: Prior to the development of a wide range of relatively low-toxic synthetic pesticides, some alkaloids, such as salts of nicotine and anabasine, were used as insecticides. Their use was limited by their high toxicity to humans.

Use as Psychoactive Drugs: Preparations of plants containing alkaloids and their extracts, and later pure alkaloids, have long been used as psychoactive substances. Cocaine, caffeine, and cathinone are stimulants of the central nervous system. Mescaline and many of indole alkaloids (such as psilocybin, dimethyltryptamine and ibogaine) have hallucinogenic effect. Morphine and codeine are strong narcotic pain killers.

There are alkaloids that do not have strong psychoactive effect themselves, but are precursors for semi-synthetic psychoactive drugs. For example, ephedrine and pseudoephedrine are used to produce methcathinone and methamphetamine. Thebaine is used in the synthesis of many painkillers such as oxycodone.

Phytotoxin

Phytotoxin refers to a substance, known as phytotoxic substances, in soils inhibitory to the growth of or poisonous to plants. Phytotoxic substances may result from human activity, or they may be produced by plant roots, by microorganisms, or by naturally-occurring chemical reactions. A good soil will protect plants from toxic concentrations of such substances by ventilating gases, by decomposing or adsorbing organic toxins, or by suppressing toxin-producing organisms. Many substances produced by plants are secondary metabolites and are the by-products of primary physiological processes. Some examples of phytotoxins exist among the plant secondary compound classes of alkaloids, terpenes, and especially phenolics, though not all such compounds are phytotoxic. They can be found as well in herbicides and substances produced by bacteria.

Toxins Produced By Plants

Alkaloids: Alkaloids are derived from amino acids, and contain nitrogen. They are medically important by interfering with components of the nervous system affecting membrane transport, protein synthesis, and enzyme activities. They generally have a bitter taste. Alkaloids usually end in -ine (caffeine, nicotine, cocaine, morphine, ephedrine).

Terpenes: Terpenes are made of water insoluble lipids, and synthesized from acetyl CoA or basic intermediates of glycolysis They often end in -ol (menthol) and make the majority of plant essential oils.

- Monoterpenes are found in gymnosperms and collect in the resin ducts and maybe released after an insect begins to feed to attract the insect's natural enemies.

- Sesquiterpenes are bitter tasting to humans and are found on glandular hairs or subdermal pigments.
- Diterpenes are contained in resin and block and deter insect feeding. Taxol, an important anticancer drug is found in this group.
- Triterpenes mimic the insect molting hormone ecdysone, disrupting molting and development and is often lethal. They are usually found in citrus fruit, and produce a bitter substance called limonoid that deters insect feeding.
- Glycosides are made of one or more sugars combined with a non-sugar like aglycone, which usually determines the level of toxicity. Cyanogenic glycosides are found in many plant seeds like cherries, apples,and plums. Cyanogenic glycosides produce cyanide and are extremely poisonous.Cardenolides have a bitter taste and influence NA+/K+ activated ATPases in human heart, they may slow or strengthen the heart rate. Saponins have lipid and water soluble components with detergent properties. Saponins form complexes with sterols and interfere with their uptake.

Phenolics: Phenolics are made of a hydroxyl group bonded to an aromatic hydrocarbon. Furanocoumarin is a phenolic and is non-toxic until activated by light. Furancoumarin blocks the transcription and repair of DNA. Tannins are another group of phenolics and they are important in tanning leather. Lignins, also a group of phenolics, is the most common compound on earth and helps conduct water in plant stems and fill spaces in the cell.

Substances Toxic to Plants

Herbicides: Herbicides usually interfere with plant growth and often imitate plant hormones.

- ACCase Inhibitors kill grasses and inhibit the first step in lipid synthesis, acetyl coA carboxylase,thus affecting cell memebrane production in the meristems. They do not affect dicots plants.
- ALS Inhibitors affect grasses and dicots by inhibiting the first step in some amino acid synthesis, acetolactate synthesis. The plants are slowly starved of theses amino acids and eventually DNA synthesis stops.
- ESPS Inhibitors affect grasses and dicots by inhibiting the first step in the synthesis of tryptophan, phenylalanine and tyrosine, enolpyruvylshikimate 3-phosphate synthase enzyme.
- Photosystem II Inhibitors reduce the electron flow from water to $NADPH^{2+}$ causing electrons to accumulate on chlorophyll molecules and excess oxidation to occur. The plant will eventually die.

- Synthetic Auxin mimics plant hormones and can affect the plant cell membrane.

Bacterial Phytotoxins

- Tabtoxin is produced by *Pseudomonas syringae pv. tabaci* that may cause toxic concentrations of ammonia to build up. This build up of ammonia causes leaf chlorosis.
- Glycopeptides are produced by a number of bacteria and have been indicated in disease development. A glycopeptide from *Corynebacterium sepedonicum* causes rapid wilt and marginal necrosis. A toxin from *Corynebacterium insidiosum* causes plugging of the plant stem interfering with water movement between cells. Amylovorin is a polysaccharide from *Erwinia amylovora* and causes wilting in rosaceous plants. A polysaccharide from *Xanthomonas campestris* obstructs water flow through phloem causing black rot in cabbage.
- Phaseolotoxin is a modified tripeptide [Nδ-(N2 - sulfodiaminophosphinyl)-ornithyl-alanyl-homoarginine] produced by certains strains of *Pseudomonas syringae* pv. *phaseolicola*, *Pseudomonas syringae* pv. *actinidiae* and strain *Pseudomonas syringae* pv. *syringae* CFBP 3388. Phaseolotoxin is a reversible inhibitor of the enzyme ornithine carbamoyltransferase (OCTase; EC 2.1.3.3), which catalyzes the formation of citrulline from ornithine and carbamoylphosphate in the arginine biosynthetic pathway. Phaseolotoxin is an effective inhibitor of OCTase activity from plant, mammalian, and bacterial sources and causes a phenotypic requirement for arginine. Additionally, phaseolotoxin inhibits the enzyme ornithine decarboxylase (EC 4.1.1.17), which is involved in the biosynthesis of polyamines.
- Rhizobiotoxine produced by *Rhizobium japonicum* causes the root nodules of some soy bean plants to become chlorotic.

Mind-Altering Plant Alkaloids

During the past 130 million years, flowering plants have colonized practically every habitat on earth, from arid deserts, boggy meadows and windswept alpine summits, to sun-baked grasslands, lush rain forests and wave-battered rocky shores. They have replaced most of the ancient ferns and seed plants that dinosaurs subsisted on, and developed a complex and fascinating relationship with insects and mammals. During these countless centuries of time, flowering plants have gradually evolved all sorts of ingenious protective devices to discourage hungry herbivorous animals. Leaves and stems have developed a variety of vicious spines and stinging hairs (trichomes). In some plants, the dense covering of silvery hairs may also provide other ecological advantages such as solar reflection and insulation in arid environments. But mechanical defences, such as spines and trichomes, are of limited value and probably would not deter all hungry herbivores, particularly the chewing and sucking insects. Therefore, plants have developed a "chemical warfare," a defence strategy based on a vast arsenal of chemicals which are toxic or distasteful to animals. According to Daniel Janzen, noted authority of the Costa Rican rain forest, developing seeds of the tropical liana Mucuna (including M. urens and M. pruriens) are nearly free from seed predators. [These vines produce the distinctive hamburger-shaped "sea beans" that drift ashore on tropical beaches throughout the world.] In addition to a dense covering of stinging trichomes, the pods of Mucuna are rich in the potentially toxic amino acid L-dopa. Just as genetic variability and time allow agricultural pests to tolerate pesticides, so can some herbivores circumvent a plant's natural chemical defences. This "predatory pressure" has resulted in the evolution of an endless array of complex plant molecules, from gums and terpenes to alkaloids and phenolic compounds. For example, in nettles (Urtica species) the sophisticated defence chemicals acetylcholine and histamine are employed in an ingenious system of "injection hairs" strategically placed throughout the plant. When we touch these plants, we may be accidental casualties in a chemical warfare between plants and herbivores that has waged through countless millennia.

One of the largest groups of chemical arsenals produced by plants are the alkaloids. Many of these metabolic by-products are derived from amino acids and include an enormous number of bitter, nitrogenous compounds. According to R.F. Raffauf, more than 10,000 different alkaloids have been discovered in species from over 300 plant families. Alkaloids often contain one or more rings of carbon atoms, usually with a nitrogen atom in the ring. The position of the nitrogen atom in the carbon ring varies with different alkaloids and with different plant families. In some alkaloids, such as mescaline, the nitrogen atom is not within a carbon ring. In fact, it is the precise position of the nitrogen atom that effects the properties of these alkaloids. Although they undoubtedly existed long before humans, some alkaloids have remarkable structural similarities with neurotransmitters in the central nervous system of humans, including dopamine, serotonin and acetylcholine. The amazing effect of these alkaloids on humans has led to the development of powerful pain-killer medications, spiritual drugs, and serious addictions by people who are ignorant of the properties of these powerful chemicals.

A Tree Protected by Ants Instead of Alkaloids

Thousands of different alkaloids have been discovered from throughout the plant kingdom, but there are some species that do not contain any of these bitter, nitrogenous compounds. One of the most remarkable examples of the "competitive disadvantage" of a tree unable to produce alkaloids is the bullhorn acacia (Acacia cornigera) of Central America. The common name of "bullhorn" refers to the large, swollen thorns (technically called stipular spines) that occur in pairs at the base of leaves, and superficially resemble the horns of a steer. In fact, the thorns of A. cornigera, and a related species A. collinsii, are so striking that they are often strung into unusual necklaces and belts. In El Salvador the horn-shaped thorns provide the legs for small ballerina seed dolls which are worn as decorative pins.

According to the famous botanist G.L. Stebbins (Flowering Plants: Evolution Above The Species Level, 1974), other species of Acacia growing in the same region of Central America (including A. chiapensis and A. macrantha) produce alkaloids, but grow much more slowly and live in drier areas than A. cornigera. Apparently these other species of Acacia pay the price for their alkaloid defence system in terms of a slower growth rate, which enables them to compete only in the drier regions where the growth of competing vegetation is also slower and less vigorous. The ability of A. cornigera to enter more moist, lush vegetation areas is apparently related to its evolution of another defence system that depends on a little hymenopteran helpmate rather than the production of bitter alkaloids.

The "swollen-thorn" acacias of Central America are truly remarkable trees. In the wild, their enlarged, hollowed-out stipular spines are occupied by fiercely biting-stinging ants that protect them from browsing herbivores and epiphytic plants that might shade them out. The swollen thorns are not galls, they are not produced in response to chemical or physical stimuli from invasive insects imbedded in their tissues. The amazing thorns are genetically-programmed structures that are formed with or without the presence of symbiotic ants. On wild trees they are inhabited by colonies of Pseudomyrmex ferruginea, very aggressive ants with a painful sting. Disturbed ants release an alarm pheromone and rush out of their thorn "barracks" in great numbers. According to Daniel Janzen (Costa Rican Natural History, 1983), livestock can apparently smell the pheromone and avoid these acacias day and night. Getting stung in the mouth and tongue is an effective deterrent to browsing on the tender foliage. In addition to protecting A. collinsii from leaf-cutting ants and other unwanted herbivores, the ants also clear away invasive seedlings around the base of the tree that might overgrow it and block out vital sunlight. According to Daniel Janzen (Smithsonian Contributions to Botany Number 13, 1974), the Central American swollen-thorn acacias lack the chemical defences of most other acacias to discourage ravaging insect predators and competition, and symbiotic ants have taken over this vital role. The acacias reward their ant helpmates with thorn "condos" to live in, carbohydrate-rich nectar from glands on the leaf stalks, and nourishing, protein-lipid morsels called Beltian bodies on the leaflet tips. There is no known function for Beltian bodies, except to provide food for symbiotic ants.

Prior to settling on a thorn acacia, the winged virgin queen ant goes on a mating flight to the highest treetop or nearby hill. Here she gets inseminated by a winged male and then hunts for an acacia in which to lay her eggs. [This "hilltopping" phenomenon also occurs in San Diego County, California, where sexually mature adult male and female butterflies fly upslope to a rendezvous point at the summit!] The queen ant cuts an entrance hole into a green thorn, hollows it out, and then deposits her eggs. Subsequent entrance holes are cut by the new generations of worker ants. All this astonishing, complicated behaviour has evolved in a tree that for some reason doesn't synthesize alkaloids.

Chile peppers in the wild may also benefit from a naturally-occurring alkaloid present in their fruits. The active ingredient causing the intense burning pain when you chew into the walls of these fruits, especially the placental region where the seeds are attached, is the alkaloid capsaicin. So potent is this alkaloid that one millionth of a drop can be detected by the human tongue. Capsaicin is not broken down during the digestion process—this is why you often get burned several hours later after dining on chile peppers. Like other alkaloids in the chemical arsenal of plants, capsaicin may serve to discourage mammalian fruit predators. Botanists believe that birds are immune to the burning sensation of capsaicin, and may serve to disperse the seeds. Capsaicin may prevent hungry mammals from devouring the fruits, so that they can be eaten by fruit-eating birds who are attracted to bright red fruits. Passing through the bird's digestive tract relatively unharmed, the small seeds are dispersed to other favourable regions. According to D. Dewitt and P. W. Bosland, there are 5 species of Capsicum peppers native to the New World: C. pubescens, C. baccatum, C. annuum, C. frutescens and C. chinense. The hottest chile peppers belong the C. chinense group, including the notorious habanero. Although this species is named "chinense," it is not from China. Actually, its centre of origin is thought to be the Amazon Basin of South America.

The Deadly Datura

There are approximately 25 different species of Datura throughout the world, including Europe, Africa, southeast Asia, Central and South America, Mexico and the United States. They are often called jimsonweed or "thornapple." The latter name refers to the spiny seed-bearing capsules. Most of the species are low, shrubby or sprawling annuals or perennials, but some tree-like forms may reach 11 metres (36 feet) in height. The tree-like forms are occasionally cultivated as ornamentals and are usually placed in the genus Brugmansia. They generally have the same characteristic trumpet-shaped flowers as Datura with striking colour variations of red, pink and yellow. Both genera belong to the Nightshade Family (Solanaceae), along with the deadly nightshade (Solanum dulcamara) , tomato (Lycopersicum esculentum), eggplant (Solanum melongena) and potato (S. tuberosum). Many species of Solanum (called nightshades) are quite poisonous to humans if eaten raw. Incidentally, the new sprouts of potato tubers contain the toxic alkaloids solanine and solanidine. Until about a century ago the tomato was thought to be poisonous because of its toxic relatives.

The common native Datura of the western United States, with rank-smelling foliage and large, white flowers up to 20 centimetres (8 inches) long is D. wrightii. Sometimes the flowers are tinged with purple and resemble oversized petunia or morning glory blossoms. It is a sprawling perennial with an enormous taproot that may extend more than 60 centimetres (2 feet) into the ground. D. wrightii has a large geographic range, including California, Utah, Arizona, New Mexico and Texas. It is often listed in older references as D. meteloides, a name that actually applies to a similar Mexican species, D. inoxia ssp. inoxia. Another species with smaller flowers, D. discolour, is native to desert washes and riverbeds of southeastern California and Arizona.

Athird species, Datura stramonium, is naturalized throughout the United States. Most older references list D. stramonium as an Old World species naturalized as an weed in North America. According to Dr. Robert Bye of the Universidad Nacional Autonoma de Mexico (personal communication, 1987), D. stramonium probably originated in the New World and migrated to the Old World during pre-Columbian time. It can readily be distinguished from D. wrightii by its more erect habit, smaller flowers and distinctive angular or prismatic calyx. It is a cosmopolitan annual weed and prolific seed-producer. A dozen viable seeds planted in fertile soil in spring may give rise to nearly 50,000 seeds by late summer, an increase of more than 400,000 percent.

Although jimsonweed is synonymous with several species of Datura, this unusual common name is actually derived from D. stramonium. In 1676 British soldiers stationed in Jamestown, Virginia became intoxicated by D. stramonium when it was inadvertently included in their salads by the regimental cooks. The episode was widely publicized and the plant culprit became known as "Jamestown weed", and later as jimsonweed.

The leaves, stem, root and fruits of Datura contain a battery of tropane alkaloids, the most potent of which are atropine, hyoscyamine and scopolamine. These alkaloids affect the central nervous system, including nerve cells of the brain and spinal cord which control many direct body functions and the behaviour of men and women. They may also affect the autonomic nervous system, which includes the regulation of internal organs, heartbeat, circulation and breathing. One autonomic response of atropine is the dilation of pupils, once considered to be a beautiful and mysterious look in Italian women. In fact, belladonna means "beautiful lady," so named because sap from the closely related belladonna plant (Atropa belladonna) was used as eye drops to dilate the pupils.

The action of tropane alkaloids at the cellular level is complex. It is summarized by R.E. Schultes and A. Hofmann in the The Botany and Chemistry of Hallucinogens (1973) and in Medical Botany by W.H. Lewis and M.P.F. Elvin-Lewis (1977). Tropane alkaloids are found in many other poisonous plants, including henbane (Hyoscyamus niger), pituri (Duboisia spp.), and mandrake (Mandragora officinarum), all of which were used extensively in witches' brews and folk medicines.

A common property of tropane alkaloids is a methylated nitrogen atom $N\text{-}CH_3$ at one end of the molecule. This chemical structure is also found in the neurotransmitter acetylcholine, which transmits impulses between nerves in the brain and neuromuscular junctions.

The anesthetic properties of tropane alkaloids may relate to their interference with acetylcholine, perhaps by competing with it at the synaptic junctions, thus blocking or inhibiting nerve impulses. It is interesting to note that the infamous tropane alkaloid, cocaine, is also a local anesthetic when injected into skin or muscle tissue. This property led to the discovery and synthesis of the more potent compound, novocain, widely used in dentistry.

Without getting into complicated anatomy and physiology, one nerve cell (neuron) connects to an adjacent neuron by a long extension called an axon. The axon branches into axonal endings, each of which attaches to the adjacent neuron at a synaptic knob filled with acetylcholine. The minute gap or synaptic cleft within this knob is only about 0.02 micrometres. As a nerve impulse (wave of depolarization or action potential) reaches this gap, acetylcholine diffuses across the synaptic cleft and activates the adjacent neuron.

Acetylcholine in the synaptic cleft is deactivated or broken down by the enzyme acetylcholinesterase, thus shutting off the action potential. Organophosphate insecticides, such as malathion and parathion, bind to active sites on this enzyme, thus preventing the normal shut down of nerve impulses and destroying the nervous control of insects. Nerve gasses developed during World War II have a similar effect on the nervous system. Gulf War soldiers carried an atropine syrette to counter the possible effects of nerve gas.

Depending on the dosages, several tropane alkaloids of Datura (when absorbed together) may have synergistic properties resulting in extreme hallucinations, delirium and death. Since the alkaloids are fat soluble they are readily absorbed through the skin and mucous membranes. Volumes have been written about the uses and properties of Datura in the Middle Ages. Most of the uses involved the consumption of potions or concoctions made from various parts of the plant.

During ancient religious rituals in India, seeds were eaten by priests to induce hallucinogenic, prophetic and oracular states. European priests apparently drank Datura for the same reason. Some authorities believe the intoxicating smoke inhaled by Greek priests over 2,000 years ago at the Oracle of Delphi was Datura. Thieves in India and Europe used Datura for centuries as "knockout drops" to rob their stupefied victims. The plant was also known in China, where a law prohibited mixing it with wine and other drinks.

In the East Indies, women fed Datura leaves to beetles, and then fed the poisonous dung to faithless lovers. Prostitutes in India added the seeds to their patron's drinks to induce sexual excitement. In fact, the use of Datura as an aphrodisiac spread throughout India, the Far East and Europe, and was an important ingredient in love potions and witches' brews. Specially prepared salves and ointments were also applied to various parts of the body. The famous seventeenth century Dutch artist, David Teniers the Younger, made several paintings of witches preparing for their demonic orgy or sabbat. The scenes frequently depicted a nude witch being anointed while she straddled a broom. According to M.J. Harner, writing in Hallucinogens and Shamanism (1973), the use of a broom or staff was undoubtedly more than a symbolic Freudian act, for it served to apply the salves to sensitive vaginal membranes.

Clay tablets from Babylonian and Assyrian ruins indicate that Datura was used medically in ancient civilizations several thousand years ago. Greek and Roman physicians used Datura mixed with opium as a sedative and general anesthetic during surgery. In fact, the use of scopolamine (one of the alkaloids in Datura) plus morphine as an effective pain reliever and sleep inducer was common practice until the nineteenth century.

One of the most interesting medical cases of the 20th century involved a German physician, Dr. Carl Gauss, and the Lutheran Church. In 1905, Dr. Gauss used extracts from Datura and morphine to induce twilight sleep treatment for women experiencing difficult child birth. The church elders denounced Dr. Gauss because the Old Testament said that women were to bring forth in pain (Genesis 3:16). The combination of scopolamine (one of the active alkaloids found in Datura) plus morphine was used for years as an effective pain reliever and sleep inducer; however, it is generally considered unsatisfactory for women in labour because of hallucinogenic side effects on the mother, and it may repress breathing of the newborn.

Datura had a number of uses among Indian tribes of the United States, as well as Mexico and South America. In fact, it is difficult to find a tribe that didn't use a species of Datura from their region in one way or another. The hallucinogenic uses involved drinking an infusion made from the crushed roots or sometimes the crushed seeds fermented in water. Some aboriginal Indians in South America gave a Datura-alcohol beverage to wives and slaves of dead warriors and chieftains. The powerful brew induced stupor before they were buried alive to accompany their dead husbands and masters on the long journey to heaven. The high priests of some tribes took Datura in order to communicate with spirits of the dead and with their gods. The brilliant red-flowered tree Datura (Brugmansia sanguinea) is used today by herbal healers called "curanderos" in several countries of South America.

Probably the best known use of Datura by several North American Indian tribes was the puberty ceremonial dances involving the drinking of a "toloache" (Datura) infusion by young boys preparing to enter manhood. The ceremonies generally involved wild erratic dancing, varied hallucinations, and finally unconsciousness.

Datura alkaloids even have modern medicinal uses. The fat solubility of tropane alkaloids is employed in a remarkable remedy for seasick mariners called Transderm Scop®. A small, medicated disc resembling a circular Band-Aid® releases small amounts of scopolamine into the bloodstream for up to 3 days. The anti-nausea effects of scopolamine may relate to its anticholinergic action on the central nervous system, which results in inhibition of the vomiting reflex.

The Jimsonweed Junkie Moth

The nocturnal blossoms and pollination ecology of Datura wrightii are among the most ingenious and unusual of all wildflowers of western North America. The entire corolla is neatly folded or pleated (plicate) and twisted (convolute) in the bud forming a compact cylinder. Each day at dusk during the summer months the buds begin to gradually unfold. Each corolla slowly unfurls and then suddenly snaps open as the intertwined lobes come lose from one another. At this instant a powerful fragrance is emitted.

Five long stamen filaments are attached to the funnelform corolla at the throat. The filaments extend as ridges down the inner corolla tube, forming 5 narrow canals to the base of the ovary where the disc-shaped nectary is located. The nectar canals can best be seen by examining a corolla tube in cross section. The curious name "revolver flower" refers to nectar canals which resemble the cylinder chambers of a 5-shot revolver. At night the nectar oozes along the length of the canals within the corolla tube.

During mid and late summer the white, fragrant blossoms are frequently visited by large nocturnal hawk moths (family Sphingidae). They are sometimes called sphinx moths because the alarm posture of the larva resembles the Egyptian sphinx. Several species of hawk moths are known to visit blossoms of Datura wrightii, but two of the most common are Manduca quinquemaculata and M. sexta. The larval forms of both are better known as tomato and tobacco hornworms.

Since Datura, tomatoes and tobacco all belong to the Nightshade family (Solanaceae), the larvae are apparently content to feed on whichever plant is available to them. The larvae are remarkably camouflaged with green markings and are difficult to spot as they rapidly devour your tomato plants. After feasting on Datura (or your tomato plants) all summer, the robust, ravenous caterpillars crawl to the ground and burrow into the soil where they undergo pupation. Unlike many other moth larvae they do not spin a cocoon. Probably every tomato gardener has unearthed the large, carmel-coloured pupa with its peculiar "jug handle" appendage, which is actually a case for the developing proboscis of the adult moth.

In warm regions there may be two generations per year: summer pupae produce adults after only a week or two; fall pupae remain in the ground until the following late spring or summer. The adult moth emerges from the pupal case and pushes out of the soil, eventually flying off into the night. After mating the gravid (pregnant) female lays her eggs on a convenient Datura plant—or on your prize-winning tomato plants while you sleep.

Adult moths have remarkable proboscides (tongues) up to 12 centimetres long (over 4 inches), long enough to reach the base of a 10-11 centimetre (4 inch) corolla tube. When the moth is not feeding the remarkable tongue rolls up into a neat, compact coil. The common white-lined sphinx moth (Hyles lineata) of the American southwest, with a medium proboscis of 3-5 centimetres (2 inches), can only reach nectar in the upper part of the nectar canals. Unlike the Manduca moths, its pupa lacks the characteristic "jug handle" appendage.

Drs. Verne and Karen Grant at the University of Texas have made detailed observations of hawk moth behaviour (Botanical Gazette 144: 280-284, 1983). The hawk moth approaches the Datura flower with its proboscis extended, lands on the blossom, crawls onto the corolla throat, and inserts its proboscis into the nectar canal. This is somewhat unusual because on most types of flowers hawk moths feed from a hovering position. While feeding they often flap their wings against the corolla throat. The proboscis becomes moist and sticky after insertion into the nectar canal, and pollen from anthers adheres to it readily. Pollen also gets on the head and other furry body parts. Thus pollen is transferred from one flower to another as the moth rubs against sticky receptive stigmas at the tips of long exerted styles. According to the Grants, hawk moths are the only flower-visiting insect with mouthparts that fit the long, slender nectar canals of Datura wrightii.

Another fascinating aspect of the Grant's research concerns a rather unconventional type of floral reward for hawk moths visiting Datura blossoms. Several intoxicating alkaloids are known to occur in Datura, but heretofore have not been correlated with pollination. Apparently Datura nectar is "spiked" with alkaloids and the hawk moths seem to like it and come back for more. Sometimes they arrive early and hover around the flowers, impatiently waiting for the blossoms to "pop" open. Outside the windows at the WAYNE'S WORD® headquarters we have observed what appear to be intoxicated moths flying erratically around D. wrightii, clumsily landing on blossoms and crashing into leaves or falling upon the ground. With a flashlight their eyes glow bright red in the darkness.

Although the coincidence of proboscis and floral lengths in hawk moths and Datura is quite remarkable, these coadapted partners have the ability to obtain nectar and set seeds in the absence of each other. According to Verne Grant (Botanical Gazette 144: 439-449, 1983), hawk moths may feed on a variety of bee flowers and hummingbird flowers, and pollen-collecting bees may visit Datura blossoms in late afternoon and early morning. The hippo staff at WAYNE'S WORD® has observed large carpenter bees (Xylocopa) and bumble bees (Bombus) on the blossoms of D. wrightii, busily collecting pollen from the anthers.

Flowers of some Datura species are also homogamous (stamens and pistils mature at the same time) and are self-compatible and partially self-pollinating. It is thus possible for one partner to migrate beyond the range of the other partner for a time. According to Verne Grant, flexibility in the pollination system permits migrational and evolutionary change. During times of general flower scarcity, hawk moths are more dependent on Datura and related sphingophilous (hawk moth-loving) blossoms because of the reliable nectar source which is unavailable to other animals.

Figure: *This beautiful orchid native to Madagascar is often called the "Darwin orchid" (Macroplectrum sesquipedale), also known as Angraecum sesquipedale. In fact, to reach the nectar of this orchid requires a 30 centimetre (eleven inch) proboscis to penetrate the long nectar spur (white arrow). Long before it was found in nature, both Charles Darwin and Alfred Wallace (founding fathers of the principles of evolution) predicted that it would be a long-tongued hawk moth.*

There are many other examples of large, white blossoms pollinated by nocturnal hawk moths, including the fragrant gardenias of French Polynesia and South Africa. In fact, to reach the nectar of a Madagascar orchid (Macroplectrum sesquipedale) requires a 30 centimetre (eleven inch) proboscis. This remarkable orchid is also known as Angraecum sesquipedale. Long before it was found in nature, both Charles Darwin and Alfred Wallace (founding fathers of the principles of evolution) predicted that it would be a hawk moth.

Although the potent alkaloids of Datura have produced untold suffering and psychedelic binges among people through countless generations, they may also be an ingenious strategy to insure repeat visits by long-tongued hawk moths through the medium of drug addiction. When the moths sober up they come straight back for more nectar. Only recently are scientists beginning to solve the mysteries of plant toxins and how they may serve as chemical defences or attractants, rather than simply nonadaptive by-products of plant evolution.

Old And New World Hallucinogenic Mushrooms

Most fungi are not deadly to humans and many are perfectly edible (and are quite delicious); however, some species are poisonous and contain potent neurotoxins. Placing a silver coin in a pan of cooking mushrooms to see if it turns black is not a reliable method of testing poisonous mushrooms. Unless you understand fungal terminology and know how to use a good taxonomic key, the staff at WAYNE'S WORD® does not encourage self indulgence on wild mushrooms. The beautiful, red, fly agaric mushroom (Amanita muscaria) is unmistakable with its bright red cap covered with white scales. It contains the toxic alkaloid, muscimol, which is derived from ibotenic acid—an amino acid. In Europe the mushrooms were reportedly left in open dishes to kill flies; however, according to some authorities, the flies are merely stunned or stupefied by the toxin, and may even regain control and fly away. Although it is poisonous to humans, there are other species of Amanita that are much more dangerous and are potentially lethal if ingested. Some of these dangerously poisonous species are death cap (A. phalloides), death angel (A. ocreata), and panther amanita (A. pantherina). Fortunately these latter deadly poisonous species are not bright red and are seldom confused with A. muscaria; however, they may be confused with other edible mushrooms by inexperienced gourmets. According to David Arora (Mushrooms Demystified, 1986), this species may be one of the most common causes of mushroom poisoning in the Pacific Northwest. David Isaak (personal communication, 2002) stated in a recent e-mail message that in the Pacific Northwest, a number of people gather "the panther" for culinary purposes; cooking it in several changes of water to remove "most" of the psychactive materials.

When ingested by humans, Amanita muscaria may produce visions and delirium, and it is perhaps one of the oldest known hallucinogens.

Recent studies suggest that this mushroom was the mysterious God-narcotic "Divine Soma" of ancient India. Thousands of years ago, Aryan conquerors who swept across India, worshiped soma, drinking it in religious ceremonies. Many hymns in the Indian Rig-Veda are devoted to Soma and describe the mushroom and its effects. According to the Rig-Veda, Soma is without leaves, seeds or branches, with a head and stalk or pillar [the structure of a mushroom]; its dazzling red skin is like the hide of the bull [the red cap]; its dress like that of a sheep, with woolly fragments remaining when the envelop bursts [the outer membranous envelop called the universal veil breaks as the stalk grows upward, leaving white remnants on the red cap]. This is a remarkably accurate description of the fly agaric mushroom (A. muscaria). There are reports of Siberian tribesmen who ingested the mushroom to get intoxicated. Since the active chemical (muscimol) passes through the body relatively unaltered, others would drink the urine from these men to get high. This way a few mushrooms could inebriate many people relatively safely and efficiently. Lapland shamans eat fly agaric mushrooms for enlightenment, and some authors have postulated that this may have given rise to the flying reindeer and the red- and white-costumed Santa Claus legends.

Apparently not everyone agrees that the "Divine Soma" is Amanita muscaria. According to Terrence McKenna (Food of the Gods, 1992), the active alkaloid in fly agaric mushrooms (muscimol) doesn't produce the psychoactive effects described in the Rig Veda and other literature. Although the identity of the Divine Soma was eloquently presented by R.G. Wasson in 1971 (Soma: Divine Mushroom of Immortality), Terrence McKenna has continued to question the effects of Amanita muscaria, and suggested other possible candidates. Based on first-hand experience with these hallucinogens, he has suggested that a psilocybin "magic mushroom," such as Stropharia cubensis, is the true Divine Soma. In fact, he also states that the use of mind-altering psilocybin mushrooms by ancient humans in Africa may have been a catalyst in the development of language and religion in primitive cultures.

Amanita muscaria was apparently one of the sacred hallucinogenic mushrooms of the Incas, Mayans and Aztecs. [Other New World psychedelic genera included Psilocybe, Paneolus, Conocybe and Stropharia]. For the Indians of Mexico, Central and South America, partaking of these mushrooms was a deeply religious experience, enabling them to communicate with their gods. Cortez reported a mushroom (resembling Amanita muscaria) being eaten during the coronation of Montezuma, and in Guatemala stone carvings dating back to 1000 BC depict curious figures with umbrella-like tops resembling the caps and stalks of an Amanita mushroom. Mushrooms are also depicted in ancient Peruvian vessels and in the Mexican Codices. One drawing shows an animal-like messenger from god offering the sacred Amanita to a ruler seated on a throne. And a fresco in a Roman Catholic Church in Plaincouralt (Indre), France depicts Adam and Eve on either side of a tree of knowledge that is unequivocally a branched Amanita mushroom. Some scholars believe that the original story of Alice's Adventures in Wonderland, where Alice speaks to a green caterpillar who is seated on a red- and white-capped mushroom, is actually the interpretation of a mushroom experience by the author, Rev. C.L. Dodgson of Christ Church College in Oxford (better known by his pen name of Lewis Carroll). Another hallucinogenic "high" that is commonly depicted in paintings and children's stories is the infamous, "politically incorrect" picture of a witch flying on a broom—the effects of a potion made from the deadly alkaloids of several solanaceous herbs, including jimsonweed (Datura stramonium).

New World Mushrooms Containing Natural LSD

According to R.E. Schultes and A. Hofmann, a number of other hallucinogenic mushrooms are used by shamans in Mexico, Central and South America, including the four genera Psilocybe, Paneolus, Conocybe and Stropharia. The exact species selected by different shamans is determined partly by personal preference and partly by the purpose of their use. Since mushrooms are often seasonal, their regional availability may also be a factor in the selection process. The Aztecs referred to these mushrooms as "teonañacatl" (flesh of the gods).

When the Spaniards conquered Mexico, they were appalled to find the natives worshipping their deities with the help of inebriating plants. Although the Spanish conquerers hated and attacked the religious use of all hallucinogens, including "peyotyl" (peyote), "ololiuqui" (morning glory), and "toloache" (Datura), they especially resented the psilocybin mushrooms "teonanacatl." In order for the Aztecs to carry on their cultural traditions without persecution, the use of some of these hallucinogens went "underground." According to R.E. Schultes (1976), the Aztecs may have tried to protect their real sacred plant (peyote) by convincing the Spaniards that all their teonanacatl were dried mushrooms, when actually many of the "mushrooms" were really the dried, shriveled crowns of a sacred peyote cactus. The Spaniards first misidentified peyote as a mushroom in the sixteenth century when they stated that the Aztec substance "teonanacatl" and peyote were the same. It is easy to see how the Spanish authorities could have mistaken the dried crowns or tops (buttons) of peyote cactus for the caps of teonanacatl mushrooms. The Aztecs may have fooled their conquerers into thinking that these religious plants were mushrooms, while the identity of one of their most spiritual and sacred plants (peyote) was a cactus.

Even in the 1900s, botanists and anthropologists concluded that teonanacatl and peyote were the same drug. This mistake was perpetuated until 1936 when hallucinogenic mushrooms were rediscovered and definitely linked to early Mexican ceremonies.

The psychoactive alkaloid in the teonanacatl mushrooms is psilocybin, a potent indole alkaloid. Psilocin, a dephosphorylated version of psilocybin, is about 10 times stronger. After ingestion by humans, psilocybin is automatically converted into psilocin. Most psilocybin-containing mushrooms have only a trace of psilocin. According to David Aurora (1986), the common psilocibin mushroom of the Pacific coast of North America, Psilocybe cyanescens has a higher concentration of natural psilocin and is appropriately named "potent psilocybe."

Although two of the most famous species of psilocybin mushrooms are Psilocybe mexicana and Stropharia cubensis, there are literally dozens of other species in the above 4 genera with similar hallucinogenic properties. In fact, Paul Stamets describes all of the species and includes colour photographs.

Like so many LBM's (Little Brown Mushrooms), they are difficult to identify unless you are familiar with mushroom structure and spore taxonomy, and have a good compound microscope at your disposal. In fact, two deadly look-alike LBM's (Galerina autumnalis and Pholiotina filaris) resemble certain species of Psilocybe. The small ring on their stems (called an annulus) and rusty brown spores (rather than black spores) are "dead" give aways to avoid these potentially lethal mushrooms.

Indole alkaloids contain the indole carbon-nitrogen ring which is also found in the fungal alkaloids ergine and psilocybin, the neurotransmitter serotonin, and the mind-altering drug LSD. These alkaloids may interfere or compete with the action of serotonin in the brain. Ergot: A Fungus Disease Of Rye That Contains LSD

One of the most amazing stories about naturally-occurring alkaloids in fungi concerns ergot (Claviceps purpurea), a fungus that infects grains of rye and related grasses. One of the psychoactive components of ergot fungus is the alkaloid ergine (d-lysergic acid amide), better known as natural LSD. The more potent synthetic LSD, (d-lysergic acid diethylamide), also known as LSD 25, is one of the most powerful psychoactive drugs known. LSD 25 was originally synthesized from natural pshchoactive alkaloids in ergot. According to Lewis and Lewis (1977), it is 4,000 times more powerful than mescaline. Natural LSD (ergine) is also found in the seeds of two species of Mexican morning glory vines which are still ingested by native Indians in an important medicinal and religious ritual.

Ergot forms a dark, compact, fungal mass called a sclerotium where the grain would normally develop. One or several of these pelletlike sclerotia can be seen in an infected grain spike, typically extending out from the bracts (glumes).

When separated from the grain spike, the sclerotia superficially resemble rat droppings (rat pellets). The sclerotia are the source of the potent alkaloids in Claviceps purpurea. In late spring, when rye plants are in bloom, the overwintering sclerotia from the previous year's crop produce stalked ascocarps resembling microscopic fungal fruiting bodies.

The head of each ascocarp contains many embedded perithecia. The perithecia contain numerous saclike asci, each with eight ascospores. The ascospores infect the young, developing grains (ovaries) of rye plants, eventually replacing them with purplish-black sclerotia. Because it produces ascospores within saclike asci, Claviceps is placed in the fungal Class Ascomycetes.

During the Middle Ages, tens of thousands of people in Europe were afflicted with ergotism, a malady characterized by gangrenous extremities, convulsions, madness and death. They ate rye bread infested with ergot fungus containing several peptide alkaloids of the ergotamine group (including ergotamine, ergosine and ergocristine) that affect blood vessels.

Since they are potent vasoconstrictors, these alkaloids can cause gangrene if ingested in sufficient dosages. Known as "St. Anthony's Fire," ergotism was a dreaded disease in Europe. Between 990 and 1129, more than 50,000 people died of this disease in France. The disease became so devastating that in 1093 in southern France the people formed an order to take care of the afflicted, and they chose St.

Anthony as their patron saint. One of the symptons of the disease was an intense burning sensation, hence the name St. Anthony's Fire. It wasn't until 1597 (500 years after the first epidemic of ergotism) that physicians finally associated this horrendous disease with the ergot on rye. Another form of ergot poisoning involves severe hallucinations and madness, caused by pschoactive alkaloids in the sclerotia.

Figure: *A dried grain spike of ryegrass*

A number of important medical discoveries have come from the study of ergot fungus and ergotism. In 1935 the alkaloid ergonovine was isolated from ergot. Since it causes strong muscular contractions, it has been used to induce labour and to control hemmorrhaging. The alkaloid ergotamine has been used extensively to relieve migraine headaches through the constriction of blood vessels. Thousands of pounds of ergot sclerotia are harvested each year from midwestern rye farms, and are used for various prescription drugs. In 1943 chemist Albert Hofmann was studying ergot fungus, whose nuclei contain lysergic acid. When he added diethylamide he produced lysergic acid diethylamide, better known as LSD. While working on this new compound, Hoffman discovered that its strong hallucinogenic effects were similar to that of natural lysergic acid alkaloids in the seeds of "ololiuqui," morning glories used by the Aztecs in their religious ceremonies.

A Hallucinogen Alkaloid Found In Seeds & Toads

Another fascinating indole alkaloid called bufotenine occurs in the seeds of Yopo or Paricα (Anadenanthera peregrina), a South American leguminous tree of the Orinoco River basin (not to be confused with the leguminous genus Adenanthera). Indians of this region prepare a powder from the ground seeds which they use as a hallucinogenic snuff. Bufotenine (5-hydroxydimethyltryptamine) is a derivative of the indole alkaloid tryptamine, which is derived from the essential amino acid tryptophan. Tryptophan is one of the 8 (9) essential dietary amino acids in humans (which we cannot synthesize), and is widely distributed in the animal kingdom. Interestingly enough, bufotenine is also present in the skin secretion of certain toads of the genus Bufo, and explains the practice of licking toads by some people.

The popular dietary supplement 5-HTP, sold at natural food stores, is 5-hydroxytryptophan. It is made from an extract of Griffonia simplicifolia seeds from coastal West Africa. 5-HTP is a metabolic precursor of serotonin and is taken as a treatment for depression and sleeping disorders. This herb may conflict with other antidepressant medications, and it would be wise to consult with a knowledgeable physician before taking it.

The Glorious Morning Glories

The morning-glory family (Convolvulaceae) contains at least 50 genera and more than 1000 species, from high-climbing vines and woody lianas of the tropical rain forest to prostrate, trailing perennials. They decorate our fences, trellises and walls with lush green foliage and colourful funnel-shaped blossoms, and form lovely green carpets of dichondra lawn. Several vines of this family provide us with valuable and nutritious root crops, including jicama and sweet potatoes. [Note: The common supermarket jicama sold in the U.S. comes from the leguminous vine Pachyrhizus erosus (Fabaceae) from Mexico and Central America.] Although the Morning-Glory Family is usually associated with climbing vines, it also includes erect herbs, shrubs and a few trees. One unusual genus includes the parasitic dodders (Cuscuta) which smother their host with masses of twining, spaghetti-like orange stems. Other morning glories have invaded cultivated fields and have become troublesome weeds. Called "bindweeds," they literally twine themselves over other plants that happen to be in their growth path. Some morning glories, including the infamous Mary's bean (Merremia discoidesperma) are excellent seed voyagers and have colonized the distant beaches of tropical islands and atolls.

Like other sacred, mind-altering plants used by the Aztecs for worshipping their gods, the use of ololiuqui was forbidden by the Spaniards. Spanish attempts to eradicate the use of these plants resulted in secrecy by the Aztecs. As in the teonanacatl mushrooms, the identity of the plants known as ololiuqui was shrouded in mystery. Because of the similar, trumpet-shaped (funnel-shaped) blossoms, the plant was thought to be Datura, a known hallucinogen still used in Mexico. Finally in the 1930s, the seeds were clearly linked to species of morning glories.

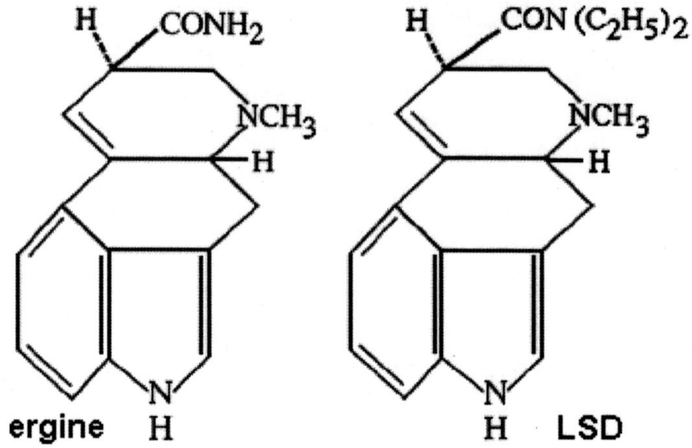

Figure: *According to R. E. Schultes and A. Hofmann (1979), the morning glory seeds contain a lysergic acid alkaloid called ergine (d-lysergic acid amide), better known as "natural" LSD. The more potent synthetic LSD is d-lysergic acid diethylamide.*

Synthetic LSD has two additional ethyl groups (C_2H_5) and is about 100 times more potent. Before it was discovered in morning glories, ergine was only known from ergot (Claviceps purpurea), a rust fungus that infects grains. Psychoactive alkaloids, such as ergine and psilocybin (from the mushrooms Psilocybe, Stropharia, Paneolus and Conocybe) contain the indole structure, a double carbon-nitrogen ring also found in the natural neurotransmitter serotonin. These alkaloids may interfere or compete with the action of serotonin in the brain, causing psychedelic visions, delusions and hallucinations.

Locoweeds: Wild and Crazy Plants

Unlike the previous mind-altering plants, the next group to be discussed are not taken as drugs plants. If ingested they can cause serious and permanent damage to the central nervous system. Nonetheless, their poisonous effects on unfortunate herbivores is quite fascinating and may lead to a better understanding of certain human genetic disorders. These plants are most commonly associated with abnormal behaviour in livestock and other range animals, and are often referred to as locoweeds. They belong to the large and diverse genus of flowering plants, Astragalus. A few western locoweeds also belong to the closely related genus, Oxytropis. Literally hundreds of different species of locoweeds grow in almost every conceivable habitat, from coastal bluffs, grasslands and deep desert canyons to sun-baked sand dunes and rocky alpine summits. In fact, Astragalus is one of the largest genera of flowering plants with approximately 2,000 different species in the northern hemisphere. Just trying to identify all the different kinds of locoweeds in the western states can be truly overwhelming. Although most North American species are poisonous, several kinds were apparently eaten by native Americans and some are actually good forage plants.

Locoweeds belong to the enormous legume family (Fabaceae), along with beans, peas, clover, alfalfa and more than 15,000 other related species. They are sometimes called milk vetches from the notion that milk secretion in goats was increased when they fed on the common Old World forage species (Astragalus cicer). Like many legumes, the leaves are typically divided into a dozen or more leaflets, and the flowers resemble small pea blossoms. Some species produce inflated seed pods that make a distinct popping sound if you step on them. Locoweeds are also called rattleweed because of the bladder-like, inflated pods, particularly when a gust of wind rattles the seeds inside. Astragalus is derived from a Greek word meaning anklebone, the plural of which means dice. Perhaps the dice connotation refers to the rattling of seeds inside papery pods, like the sound of dice in a thrower. In anatomy, the astragalus or talus is one of seven bones in the ankle joint. Anklebones were apparently used for dice by ancient Greeks, and to this day, veteran crapshooters in Las Vegas refer to dice as "bones." Since adult astragalus bones are a little too large for dice, some of the smaller, cuboidal or cuneiform anklebones were probably used.

Many species of locoweeds native to the western United States are known to be poisonous to livestock. The actual mechanism of locoweed poisoning may involve: (1) Toxic levels of selenium absorbed from the soil, (2) nitrogen-containing sugar compounds called nitroglycosides, and (3) potent swainsonine alkaloids causing a condition called "locoism." According to P.R. Cheeke and L.R. Shull, of the 372 species of Astragalus in North America, about 25 species contain toxic levels of selenium, 263 contain poisonous nitroglycosides, and 13 cause locoism. Some of the poisonous species may be placed in more than one of the above three categories. Bee keepers in Nevada have reported serious honey bee losses after working the flowers of spotted locoweed (Astragalus lentiginosus). The toxicity of Astragalus to bees is unusual because relatively few plants produce nectar or pollen which is poisonous to honey bees. The notoriously toxic "diablo locoweed" (A. oxyphysus) is well-known in California and some ranchers have attempted to eradicate it from their property.

Some species of Astragalus accumulate toxic levels of selenium and often have a peculiar malodorous foliage. In fact, specimens of A. bisulcatus and the desert species A. crotalariae can produce the dominant scent in herbarium cabinets for many years. Locoweed indicators of selenium-rich soils are sometimes referred to as "poison vetches," and may contain selenium levels of several hundred to 10,000 ppm (parts per million). Several names have been applied to selenium poisoning in cattle, including "blind staggers" and "alkali disease." In severe cases, the animals become lame and emaciated, and often fall into a kneeling position from which they are unable to rise. Toxic levels of selenium also occur in other wild plants of western North America, such as species of Stanleya, colourful shrubby perennials in the mustard family (Brassicaceae). Selenium poisoning of fish and waterfowl from has been well documented in California's Central Valley. In the human body, the trace element selenium is needed for the function of certain enzymes, and may serve as a cellular antioxidant thought to suppress certain tumors (M.A. Weiner and J.A. Weiner, Herbs That Heal, 1994). Natural sources of selenium come from the allium vegetables, including onions and garlic.

The majority of locoweed species poisonous to livestock contain toxic nitroglycosides. The extreme toxicity of nitroglycosides, such as miserotoxin, in livestock is caused by the production 3-nitropropionic acid (NPA) through a complex metabolic pathway involving enzymes of rumen microbes and the liver (W. Majak and M.A. Pass, Toxicants of Plant Origin , 1989). Another toxic reaction involves the oxidation of hemoglobin by nitrites produced by the hydrolysis of miserotoxin. The oxidized hemoglobin (called methemoglobin) is incapable of carrying oxygen. NPA is lethal to animals because it inhibits the vital enzyme succinate dehydrogenase inside mitochondria, thus blocking ATP synthesis and ultimately causing cellular death (M.A. Pass, Plant-Associated Toxins, 1994). In some animals with different gastrointestinal microbes, such as rats and rabbits, the miserotoxin is not converted into lethal NPA; however, studies indicate that death in rabbits is due to nitrite poisoning (W. Majak and M.A. Pass, 1989). Symptoms of acute livestock poisoning involve general weakness and loss of neural control, convulsions, blindness, coma, and death. More than 200 species of Astragalus in North America contain nitroglycosides, including A. cibaria, A. falcatus, and A. miser. The latter species has also been known to cause high mortality in foraging honey bees.

One of the most destructive types of livestock poisoning by locoweed ingestion is called "loco disease" or locoism. It is caused by at least two very potent indolizidine alkaloids, swainsonine and swainsonine-N-oxide, and impels horses and cattle to act in a wild and crazy manner. Some authorities only apply the term "locoweed" to species of Astragalus and Oxytropis containing these alkaloids, such as A. lentiginosus, A. mollissimus, A. wootonii, O. lambertii, and O. sericea. Swainsonine was originally isolated from the Australian darling pea (Swainsona) and more recently from the spotted locoweed (Astragalus lentiginosus) (R.J. Molyneux and L.F. James, Loco Intoxication: Indolizidine Alkaloids of Spotted Locoweed, 1981).

Consumption of locoweeds for two weeks to a month is necessary before obvious signs of poisoning are evident (L.F. James and K.E. Painter, Locoweed Poisoning in Livestock, 1989). Animals suffering from locoism may exhibit depression, a staggering gait, and general muscular incoordination. They may withdraw from other animals and become solitary. The final stages are characterized by difficulty in eating or drinking, paralysis and death. Afflicted animals often become nervous, or easily aggravated, and exhibit a complete loss of depth perception. There are reports of animals being injured or killed by running through fences or falling into ponds and streams. When plenty of other forage is available, animals tend to avoid locoweed; however, poisoned animals apparently acquire a taste for it and actually seek it out (P.R. Cheeke and L.R. Shull, 1985).

The cause of locoism at the cellular level is very complex. The swainsonine alkaloids inhibit or tie up the key enzyme mannosidase resulting in the accumulation of mannose sugar in nerve cells and irreparable damage to brain tissue (P.R. Dorling, C.R. Huxtable and P. Vogel, Neuropathology and Applied Neurobiology, 1978). This condition is remarkably similar to a genetic deficiency of the same vital enzyme in humans called mannosidosis (P.A. Ockermann, Mannosidosis, 1973). Mannosidosis is a genetic disorder called a lysosomal storage disease, in which cells of the central nervous system become filled with cytoplasmic vacuoles of mannose due to the lack of the vital enzyme mannosidase that is essential in breaking down mannose. The actual vacuoles are swollen organelles called lysosomes where the enzymatic breakdown process normally occurs. Lysosomal storage diseases, such as mannosidosis, are often caused by recessive genes and result in paralysis and death within a few years following birth. Perhaps one of the better known storage diseases is Tay Sachs Disease, in which nerve cells fill up with a lipid called ganglioside or GM2 because they lack the vital enzyme HEX A needed to break down GM2. Studies of the biochemical effects of locoweed poisoning on the central nervous system of cattle may lead to a better understanding of these tragic human neurological disorders.

Native American Indians were apparently aware of the poisonous properties of some locoweeds and avoided them. Other drug plants are occasionally called locoweeds, such as the hallucinogenic jimsonweed (Datura), and some of these were commonly utilized by Indians during ceremonial rituals. In the southwest, Cahuilla Indians utilized the pods of at least one unknown Astragalus locoweed (J.L. Bean and K.S. Saubel, Temalpakh: Cahuilla Indian Knowledge and Usage of Plants, 1972). The pods were pounded up and mixed with beans and other foods, perhaps as a spice or flavour enhancer. An Arizona locoweed (A. ceramicus), with beautiful brown-mottled pods, produces sweet, edible rootstocks which are reportedly eaten in spring by Hopi children. Another locoweed of the prairie and plains states (A. crassicarpus) produces edible fleshy pods (K. Kindscher, Edible Wild Plants of the Prairie: An Ethnobotanical Guide, 1987). It is known locally as ground plum, Indian pea, or buffalo bean, and the pods were eaten, raw or cooked, by Indians and white settlers. The perennial roots of three species, including A. canadensis, A. caryocarpus, and A. pictus-filifolius, were eaten raw or cooked by Blackfoot Indians of Montana and Hopi of Arizona. Not all locoweeds are poisonous to people; however, because of the difficulty in identifying some species, it would definitely NOT be advisable to try any of them in gourmet dishes. It would actually be easier to select edible mushrooms from the deadly species, especially with all of the pictorial mushroom guides available today.

The Mescal Bean and Peyote Cactus

Probably the most famous New World hallucinogenic plant is peyote (Lophophora williamsii), a small, spineless cactus native to the Rio Grande valley of Texas and the northern and central parts of the Mexican plateau region. Another species (L. diffusa) is native to the Mexican state of Queretaro. The rounded, gray-green stem crown (top) is radially-divided into sections, each bearing a small meristematic region (called an areole) from which arises a tuft of hairs. The crown tapers into a thick carrot-like root that extends into the ground.

Called "peyotyl" by the Aztecs, this was truly a sacred plant used in religious ceremonies where they communicated with their gods. The spiritual, religious usage of peyote persists to this day by the Tarahumara, Huichol and other Mexican Indians, as well as by members of the Native American Church in the United States and western Canada. The Indians cut off the crowns and sun-dry them into brown, mushroom-shaped "mescal buttons" that last for long periods and can be shipped to distant places for use. After the crown (button) dries, the soft, fleshy tissue is reduced in volume. For this reason the tufts of hairs appear much larger and occupy a larger proportion of the button. The Aztec word peyotyl means "caterpillar cocoon," referring to the white, woolly tufts of hairs. When the top is severed, the plant often resprouts with new crowns so that many-crowned peyote plants are common. Except for the woolly hairs, the dried, peyote buttons superficially resemble the cap of a dried mushroom.

Spanish conquerers tried to ban the use of peyote by native Indians. By 1720 the eating of peyote was prohibited throughout Mexico, but peyote was so strongly rooted in native Indian lore that its use actually spread to other tribes. The Spanish thought the peyote buttons used by Aztecs were mushrooms, and it wasn't until the 1930s that researchers finally discovered that the Aztecs were actually using both peyote (peyotyl) and psilocybin mushrooms (teonanacatl) in their religious ceremonies.

The peyote cactus contains more than 50 different alkaloids, but the most active hallucinogen is mescaline. The chemistry, botany and history of peyote is discussed in a fascinating book by E.F. Anderson (Peyote: The Divine Cactus, 1976). Mescaline has a chemical structure similar to the brain neurotransmitter dopamine. It is also structurally similar to the neurohormone norepinephrine (noradrenalin) and to the stimulant amphetamine. In the peyote cactus, mescaline is formed in a complex pathway from the amino acid tyrosine. A similar pathway in humans produces epinephrine (adrenalin) and its demethylated precursor norepinephrine from tyrosine. Dopamine and its precursor L-dopa are also derived from a tyrosine pathway. The following illustration shows the remarkable molecular similarity between mescaline and dopamine:

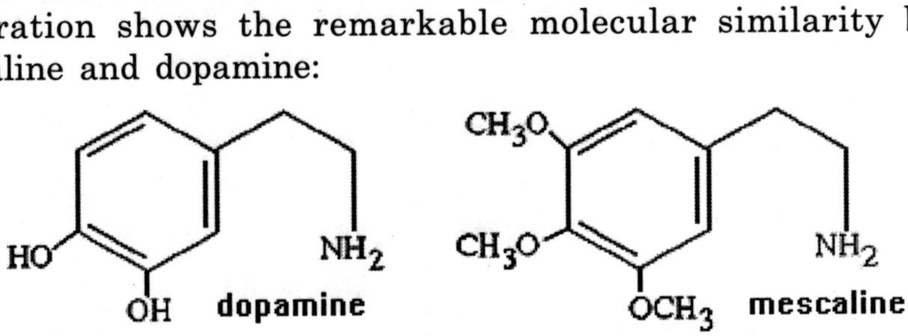

Mescaline also occurs in several other cactus species, including the commonly cultivated, night-blooming, South American San Pedro cactus (Trichocereus pachanoi). In the Andes of Peru, Ecuador and Bolivia the natives call this tall, columnar cactus "aguacolla" or "giganton." An intoxicating drink called "cimora" is made from the boiled stems. According to R.E. Schultes (1976), the drink may be spiked with other potent hallucinogens, including species of Brugmansia (Datura). The cut stems are sometimes seen in market places of the northern Peruvian Andes in neatly stacked piles. There are about 25 species of Trichocereus native to South America. They are typically large, tree-like cacti with cylindrical, ribbed stems and large, nocturnal, white blossoms. Other species within this fascinating genus may also contain mescaline.

Contrary to popular rumors, mescaline is not found in the bright red, poisonous seeds of the beautiful, drought-resistant shrub called mescal bean (Sophora secundiflora); however, they do inhabit a similar range in the arid lands of the southwestern United States and Mexico. Mescaline is also not related to the highly intoxicating beverage called "mescal" or "mezcal," made from the fermented and distilled juices of several North American species of Agave, including A. americana and A. atrovirens. Incidentally, the fermented juice is called pulque, and the highly-alcoholic distilled products include mezcal and tequila. The seeds of mescal bean contain another potent and dangerously poisonous alkaloid, cytisine, which was ingested by some North American Indian tribes in a vision-seeking "Red Bean Dance" prior to the widespread use of peyote. According to R.E. Schultes (1976) mescal beans have been discovered in Indian sites dating before AD 1000, and from one site dating back to 1500 BC. In fact, to this day the leader of the peyote ceremony in some of these tribes (called the "roadman") wears a necklace made from bright red mescal beans. [For biology students, cytisine is not to be confused with the pyrimidine base cytosine found in DNA and RNA.]

6

Cyanogenic Glycosides

Cyanogenic glycosides are present in a number of food plants and seeds. Hydrogen cyanide is released from the cyanogenic glycosides when fresh plant material is macerated as in chewing, which allows enzymes and cyanogenic glycosides to come together, releasing hydrogen cyanide. Cassava, an important source of carbohydrate for people in Africa and South America, is detoxified by chopping and grinding in running water prior to preparation.

Cyanide is one of the most potent, rapidly acting, poisons known. Cyanides inhibit the oxidative processes of cells causing them to die very quickly. Because the body rapidly detoxifies cyanide, an adult human can withstand 50-60 ppm for an hour without serious consequences. However, exposure to concentrations of 200-500 ppm for 30 minutes is usually fatal. Aside from death, acute cyanide toxicity at small doses can cause headache, tightness in throat and chest, and muscle weakness. The effects of chronic (long-term) exposure to cyanide are less well known.

***Table:** Food Sources of Cyanogenic Glycosides and Amount of hydrogen cyanide (HCN) Produced*

Plant	***mg/100 g HCN***	***Glucoside***
Bitter almonds	250	Amygdalin
Cassava root	53	Linamarin
Whole sorghum	250	Dhurrin
Lima bean	10-312	Linamarin

Glycoside

In chemistry, a glycoside is a molecule in which a sugar is bound to another functional group via a glycosidic bond. Glycosides play numerous important roles in living organisms. Many plants store chemicals in the form of inactive glycosides. These can be activated by enzyme hydrolysis, which causes the sugar part to be broken off, making the chemical available for use. Many such plant glycosides are used as medications. In animals and humans, poisons are often bound to sugar molecules as part of their elimination from the body.

In formal terms, a glycoside is any molecule in which a sugar group is bonded through its anomeric carbon to another group via a glycosidic bond. Glycosides can be linked by an O- (an *O-glycoside*), N- (a *glycosylamine*), S-(a *thioglycoside*), or C- (a *C-glycoside*) glycosidic bond. According to the IUPAC the name "*C*-glycoside" is a misnomer, the preferred term is "*C*-glycosyl compound". The given definition is the one used by IUPAC, which recommends the Haworth projection to correctly assign stereochemical configurations. Many authors require in addition that the sugar be bonded to a *non-sugar* for the molecule to qualify as a glycoside, thus excluding polysaccharides. The sugar group is then known as the glycone and the non-sugar group as the aglycone or genin part of the glycoside. The glycone can consist of a single sugar group (monosaccharide) or several sugar groups (oligosaccharide).

The first glycoside ever identified was amygdalin, by the French chemists Pierre Robiquet and Antoine Boutron-Charlard, in 1830.

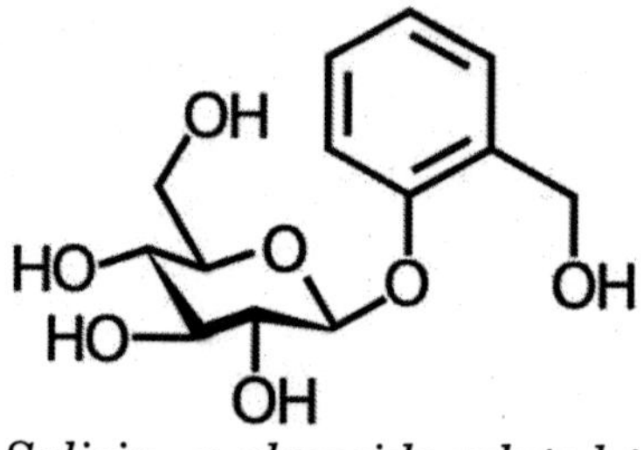

Figure: *Salicin, a glycoside related to aspirin*

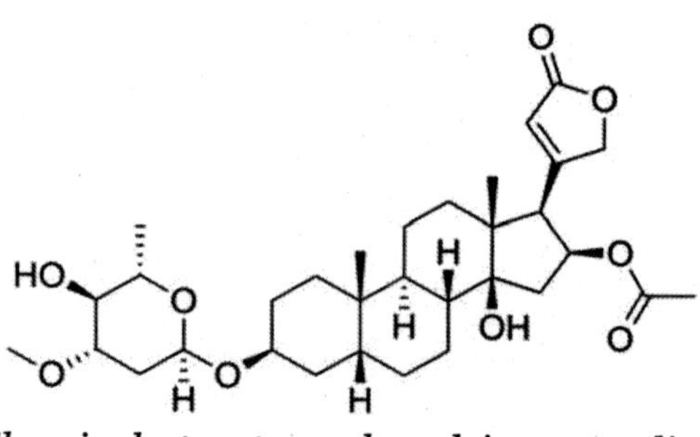

Figure: *Chemical structure oleandrin, a cardiac glycoside*

Related Compounds

Molecules containing an N-glycosidic bond are known as glycosylamines and are not discussed in this article. (Many authors in biochemistry call these compounds *N-glycosides* and group them with the glycosides; this is considered a misnomer and discouraged by IUPAC.) Glycosylamines and glycosides are grouped together as glycoconjugates; other glycoconjugates are glycoproteins, glycopeptides, peptidoglycans, glycolipids, and lipopolysaccharides.

Chemistry

Much of the chemistry of glycosides is explained in the article on glycosidic bonds. For example, the glycone and aglycone portions can be chemically separated by hydrolysis in the presence of acid and can be hydrolyzed by alkali. There are also numerous enzymes that can form and break glycosidic bonds. The most important cleavage enzymes are the glycoside hydrolases, and the most important synthetic enzymes in nature are glycosyltransferases. Genetically altered enzymes termed glycosynthases have been developed that can form glycosidic bonds in excellent yield.

There are a great many ways to chemically synthesize glycosidic bonds. Fischer glycosidation refers to the synthesis of glycosides by the reaction of unprotected monosaccharides with alcohols (usually as solvent) in the presence of a strong acid catalyst. The Koenigs-Knorr reaction is the condensation of glycosyl halides and alcohols in the presence of metal salts such as silver carbonate or mercuric oxide.

Classification

Glycosides can be classified by the glycone, by the type of glycosidic bond, and by the aglycone.

By Blycone/presence of Sugar

If the glycone group of a glycoside is glucose, then the molecule is a glucoside; if it is fructose, then the molecule is a fructoside; if it is glucuronic acid, then the molecule is a glucuronide; etc. In the body, toxic substances are often bonded to glucuronic acid to increase their water solubility; the resulting glucuronides are then excreted.

By Type of Glycosidic Bond

Depending on whether the glycosidic bond lies "below" or "above" the plane of the cyclic sugar molecule, glycosides are classified as α-glycosides or β-glycosides. Some enzymes such as α-amylase can only hydrolyze α-linkages; others, such as emulsin, can only affect β-linkages.

There are four type of linkages present between glycone and aglycone:

- C-linkage/glycosidic bond, "nonhydrolysable by acids or enzymes"
- O-linkage/glycosidic bond
- N-linkage/glycosidic bond
- S-linkage/glycosidic bond

By Aglycone

Glycosides are also classified according to the chemical nature of the aglycone. For purposes of biochemistry and pharmacology, this is the most useful classification.

Alcoholic Glycosides

An example of an alcoholic glycoside is salicin, which is found in the genus *salix.* Salicin is converted in the body into salicylic acid, which is closely related to aspirin and has analgesic, antipyretic, and antiinflammatory effects.

Anthraquinone Glycosides

These glycosides contain an aglycone group that is a derivative of anthraquinone. They have a laxative effect. They are mainly found in dicot plants except the Liliaceae family which are monocots. They are present in senna, rhubarb and *Aloe* species. Antron and anthranol are reduced forms of anthraquinone.

Coumarin Glycosides

Here the aglycone is coumarin or a derivative. An example is apterin which is reported to dilate the coronary arteries as well as block calcium channels. Other coumarin glycosides are obtained from dried leaves of *Psoralea corylifolia.*

Chromone Glycosides

In this case, the aglycone is benzo-gamma-pyrone.

Cyanogenic Glycosides

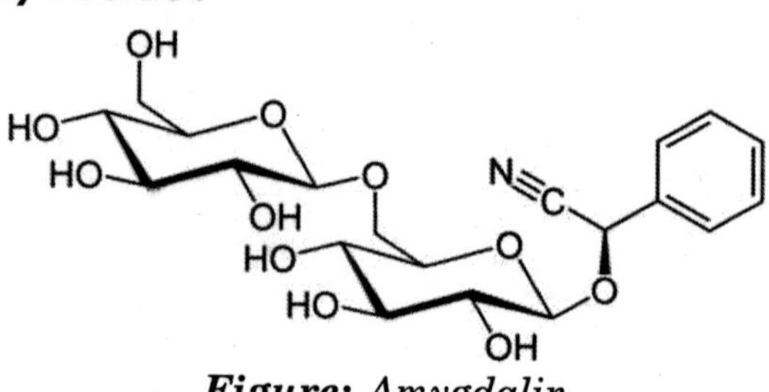

Figure: *Amygdalin*

In this case, the aglycone contains a cyanide group. All of these plants have these glycosides stored in the vacuole, but, if the plant is attacked, they are released and become activated by enzymes in the cytoplasm. These remove the sugar part of the molecule and release toxic hydrogen cyanide. Storing them in inactive forms in the cytoplasm prevents them from damaging the plant under normal conditions.

An example of these is amygdalin from almonds. They can also be found in the fruits (and wilting leaves) of the rose family (including cherries, apples, plums, almonds, peaches, apricots, raspberries, and crabapples). Cassava, an important food plant in Africa and South America, contains cyanogenic glycosides and, therefore, has to be washed and ground under running water prior to consumption. Sorghum (*Sorghum bicolour*) expresses cyanogenic glycosides in its roots and, thus, is resistant to pests such as rootworms (Diabrotica spp.) that plague its cousin maize (*Zea mays* L.). It was once thought that cyanogenic glycosides might have anti-cancer properties, but this idea was disproven. A recent study may also show that increasing CO_2 levels may result in much higher levels of cyanogenic glycoside production in sorghum and cassava plants, making them highly toxic and inconsumable. A doubling of CO_2 concentration was found to double the concentration of cyanogenic glycosides in the leaves.

Dhurrin, linamarin, lotaustralin, and prunasin are also classified as cyanogenic glycosides.

Flavonoid Glycosides

Here the aglycone is a flavonoid. Examples of this large group of glycosides include:

- Hesperidin (aglycone: Hesperetin, glycone: Rutinose)
- Naringin (aglycone: Naringenin, glycone: Rutinose)
- Rutin (aglycone: Quercetin, glycone: Rutinose)
- Quercitrin (aglycone: Quercetin, glycone: Rhamnose)

Among the important effects of flavonoids are their antioxidant effect. They are also known to decrease capillary fragility.

Phenolic Glycosides (simple)

Here the aglycone is a simple phenolic structure. An example is arbutin found in the Common Bearberry *Arctostaphylos uva-ursi*. It has a urinary antiseptic effect.

Saponins

These compounds give a permanent froth when shaken with water. They also cause hemolysis of red blood cells. Saponin glycosides are found in liquorice. Their medicinal value is due to their expectorant, and corticoid and anti-inflammatory effects. Steroid saponins, for example, in Dioscorea wild yam the sapogenin diosgenin—in form of its glycoside dioscin—is an important starting material for production of semi-synthetic glucocorticoids and other steroid hormones such as progesterone. The ginsenosides are triterpene glycosides and Ginseng saponins from *Panax Ginseng* C. A. Meyer, (Chinese ginseng) and *Panax quinquefolius* (American Ginseng). In general, the use of the term saponin in organic chemistry is discouraged, because many plant constituents can produce foam, and many triterpene-glycosides are amphipolar under certain conditions, acting as a surfactant. More modern uses of saponins in biotechnology are as adjuvants in vaccines: Quil A and its derivative QS-21, isolated from the bark of Quillaja saponaria Molina, to stimulate both the Th1 immune response and the production of cytotoxic T-lymphocytes (CTLs) against exogenous antigens make them ideal for use in subunit vaccines and vaccines directed against intracellular pathogens as well as for therapeutic cancer vaccines but with the aforementioned side-effects of hemolysis.

Steroidal Glycosides or Cardiac Glycosides

Here the aglycone part is a steroidal nucleus. These glycosides are found in the plant genera *Digitalis*, *Scilla*, and *Strophanthus*. They are used in the treatment of heart diseases, e.g., congestive heart failure (historically as now recognised does not improve survivability; other agents are now preferred) and arrhythmia.

Steviol Glycosides

These sweet glycosides found in the stevia plant *Stevia rebaudiana* Bertoni have 40-300 times the sweetness of sucrose. The two primary glycosides, stevioside and rebaudioside A, are used as natural sweeteners in many countries. These glycosides have steviol as the aglycone part. Glucose or rhamnose-glucose combinations are bound to the ends of the aglycone to form the different compounds.

Thioglycosides

As the name implies (q.v. thio-), these compounds contain sulfur. Examples include sinigrin, found in black mustard, and sinalbin, found in white mustard.

Reactive Oxygen Species

Reactive oxygen species (ROS) are chemically reactive molecules containing oxygen. Examples include oxygen ions and peroxides. ROS are formed as a natural byproduct of the normal metabolism of oxygen and have important roles in cell signaling and homeostasis. However, during times of environmental stress (e.g., UV or heat exposure), ROS levels can increase dramatically. This may result in significant damage to cell structures. Cumulatively, this is known as oxidative stress. ROS are also generated by exogenous sources such as ionizing radiation.

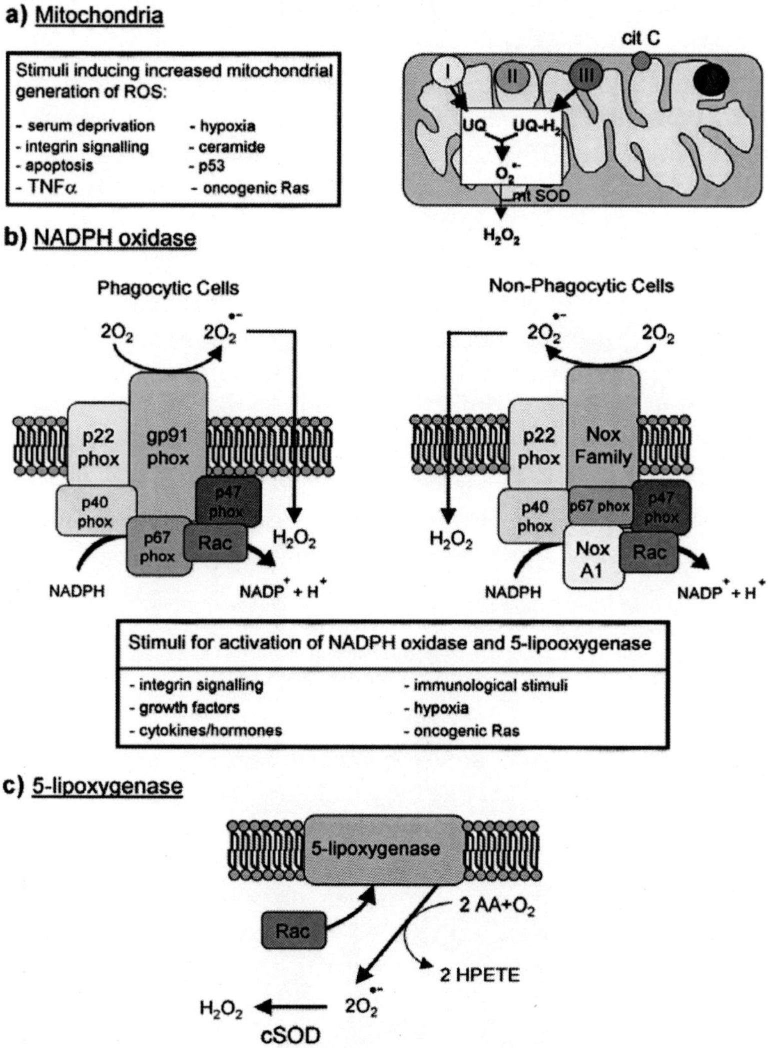

***Figure:** Major cellular sources of ROS in living non-photosynthetic cells.*

Damaging Effects

Normally, cells defend themselves against ROS damage with enzymes such as alpha-1-microglobulin, superoxide dismutases, catalases, lactoperoxidases, glutathione peroxidases and peroxiredoxins. Small molecule antioxidants such as ascorbic acid (vitamin C), tocopherol (vitamin E), uric acid, and glutathione also play important roles as cellular antioxidants. In a similar manner, polyphenol antioxidants assist in preventing ROS damage by scavenging free radicals. In contrast, the antioxidant ability of the extracellular space is less - e.g., the most important plasma antioxidant in humans is uric acid.

Effects of ROS on cell metabolism are well documented in a variety of species. These include not only roles in apoptosis (programmed cell death) but also positive effects such as the induction of host defencegenes and mobilisation of ion transport systems. This implicates them in control of cellular function. In particular, platelets involved in wound repair and blood homeostasis release ROS to recruit additional platelets to sites of injury. These also provide a link to the adaptive immune system via the recruitment of leukocytes.

Reactive oxygen species are implicated in cellular activity to a variety of inflammatory responses including cardiovascular disease. They may also be involved in hearing impairment via cochlear damage induced by elevated sound levels, in ototoxicity of drugs such as cisplatin, and in congenital deafness in both animals and humans. ROS are also implicated in mediation of apoptosis or programmed cell death and ischaemic injury. Specific examples include stroke and heart attack.

In general, harmful effects of reactive oxygen species on the cell are most often:

1. damage of DNA
2. oxidations of polyunsaturated fatty acids in lipids (lipid peroxidation)
3. oxidations of amino acids in proteins
4. oxidatively inactivate specific enzymes by oxidation of co-factors

Pathogen Response

When a plant recognizes an attacking pathogen, one of the first induced reactions is to rapidly produce superoxide or hydrogen peroxide to strengthen the cell wall. This prevents the spread of the pathogen to other parts of the plant, essentially forming a net around the pathogen to restrict movement and reproduction. In the mammalian host, ROS is induced as an antimicrobial defence. To highlight the importance of this defence, individuals with chronic granulomatous disease who have deficiencies in generating ROS, are highly susceptible to infection by a broad range of microbes including *Salmonella enterica*, *Staphylococcus aureus*, *Serratia marcescens*, and *Aspergillus* spp. The exact manner in which ROS defends the host from invading microbe is not fully understood. One of the more likely modes of defence is damage to microbial DNA. Studies using Salmonella demonstrated that DNA repair mechanisms were required to resist killing by ROS. More recently, a role for ROS in antiviral defence mechanisms has been demonstrated via Rig-like helicase-1 and mitochondiral antiviral signaling protein. Increased levels of ROS potentiate signaling through this mitochondia-associated antiviral receptor to activate interferon regulatory factor (IRF)-3, IRF-7, and nuclear facto- kappaB (NFκB), resulting in an antiviral state. Respiratory epithelial cells were recently demonstrated to induce mitrochondrial ROS in response to influenza infection. This induction of ROS lead to the induction of type III interferon and the induction of an antiviral state, limiting viral replication.

Oxidative Damage

In aerobic organisms the energy needed to fuel biological functions is produced in the mitochondria via the electron transport chain. In addition to energy, reactive oxygen species (ROS) with the potential to cause cellular damage are produced. ROS can damage DNA, RNA, and proteins, which, in theory, contributes to the physiology of ageing.

ROS are produced as a normal product of cellular metabolism. In particular, one major contributor to oxidative damage is hydrogen peroxide (H_2O_2), which is converted from superoxide that leaks from the mitochondria. Catalase and superoxide dismutase ameliorate the damaging effects of hydrogen peroxide and superoxide, respectively, by converting these compounds into oxygen and hydrogen peroxide (which is later converted to water), resulting in the production of benign molecules. However, this conversion is not 100% efficient, and residual peroxides persist in the cell. While ROS are produced as a product of normal cellular functioning, excessive amounts can cause deleterious effects. Memory capabilities decline with age, evident in human degenerative diseases such as Alzheimer's disease, which is accompanied by an accumulation of oxidative damage. Current studies demonstrate that the accumulation of ROS can decrease an organism's fitness because oxidative damage is a contributor to senescence. In particular, the accumulation of oxidative damage may lead to cognitive dysfunction, as demonstrated in a study in which old rats were given mitochondrial metabolites and then given cognitive tests. Results showed that the rats performed better after receiving the metabolites, suggesting that the metabolites reduced oxidative damage and improved mitochondrial function. Accumulating oxidative damage can then affect the efficiency of mitochondria and further increase the rate of ROS production. The accumulation of oxidative damage and its implications for aging depends on the particular tissue type where the damage is occurring. Additional experimental results suggest that oxidative damage is responsible for age-related decline in brain functioning. Older gerbils were found to have higher levels of oxidized protein in comparison to younger gerbils. Treatment of old and young mice with a spin trapping compound caused a decrease in the level of oxidized proteins in older gerbils but did not have an effect on younger gerbils. In addition, older gerbils performed cognitive tasks better during treatment but ceased functional capacity when treatment was discontinued, causing oxidized protein levels to increase. This led researchers to conclude that oxidation of cellular proteins is potentially important for brain function.

Exogenous ROS

Exogenous ROS can be produced from pollutants, tobacco, smoke, drugs, xenobiotics, or radiation.

Ionizing radiation can generate damaging intermediates through the interaction with water, a process termed radiolysis. Since water comprises 55-60% of the human body, the probability of radiolysis is quite high under the presence of ionizing radiation. In the process, water loses an electron and become highly reactive. Then through a three-step chain reaction, water is sequentially converted to hydroxyl radical (-OH), hydrogen peroxide (H_2O_2), superoxide radical (O_2-) and ultimately oxygen (O_2). The hydroxyl radical is extremely reactive that immediately removes electrons from any molecule in its path, turning that molecule into a free radical and so propagating a chain reaction. But hydrogen peroxide is actually more damaging to DNA than hydroxyl radical since the lower reactivity of hydrogen peroxide provides enough time for the molecule to travel into the nucleus of the cell, subsequently wreaking havoc on macromolecules such as DNA.

Endogenous ROS

ROS are produced intracellularly through multiple mechanisms and depending on the cell and tissue types, the major sources being the "professional" producers of ROS NADPH oxidase (NOX) complexes (7 distinct isoforms) in cell membranes, mitochondria, peroxisomes, and endoplasmic reticulum. Mitochondria convert energy for the cell into a usable form, adenosine triphosphate (ATP). The process in which ATP is produced, called oxidative phosphorylation, involves the transport of protons (hydrogen ions) across the inner mitochondrial membrane by means of the electron transport chain. In the electron transport chain, electrons are passed through a series of proteins via oxidation-reduction reactions, with each acceptor protein along the chain having a greater reduction potential than the previous. The last destination for an electron along this chain is an oxygen molecule. In normal conditions, the oxygen is reduced to produce water; however, in about 0.1–2% of electrons passing through the chain (this number derives from studies in isolated mitochondria, though the exact rate in live organisms is yet to be fully agreed upon), oxygen is instead prematurely and incompletely reduced to give the superoxide radical ($\cdot O_2^-$), most well documented for Complex I and Complex III. Superoxide is not particularly reactive by itself, but can inactivate specific enzymes or initiate lipid peroxidation in its protonated form, hydroperoxyl HO_2 $\cdot$. The pKa of hydroperoxyl is 4.8. Thus, at physiological pH, the majority will exist as superoxide.

If too much damage is present in mitochondria, a cell undergoes apoptosis or programmed cell death. Bcl-2 proteins are layered on the surface of the mitochondria, detect damage, and activate a class of proteins called Bax, which punch holes in the mitochondrial membrane, causing cytochrome C to leak out. This cytochrome C binds to Apaf-1, or apoptotic protease activating factor-1, which is free-floating in the cell's cytoplasm. Using energy from the ATPs in the mitochondrion, the Apaf-1 and cytochrome C bind together to form apoptosomes. The apoptosomes bind to and activate caspase-9, another free-floating protein. The caspase-9 then cleaves the proteins of the mitochondrial membrane, causing it to break down and start a chain reaction of protein denaturation and eventually phagocytosis of the cell.

Superoxide Dismutase

Superoxide dismutases (SOD, EC 1.15.1.1) are enzymes that catalyze the dismutation of superoxide (O_2^-) into oxygen and hydrogen peroxide. Thus, they are an important antioxidant defence in nearly all cells exposed to oxygen. One anomaly is *Lactobacillus plantarum* and related lactobacilli, which use a different mechanism.

In mammals and most chordates, three forms of superoxide dismutase are present. SOD1 is located primarily in the cytoplasm, SOD2 in the mitochondria and SOD3 is extracellular. The first is a dimer (consists of two units), while the others are tetramers (four subunits). SOD1 and SOD3 contain copper and zinc ions, while SOD2 has a manganese ion in its reactive centre. The genes are located on chromosomes 21, 6, and 4, respectively.

Reaction

The SOD-catalysed dismutation of superoxide may be written with the following half-reactions :

- $M^{(n+1)+}$-SOD + O_2^- $\rightarrow$ M^{n+}-SOD + O_2
- M^{n+}-SOD + O_2^- + $2H^+$ $\rightarrow$ $M^{(n+1)+}$-SOD + H_2O_2.

where M = Cu (n=1) ; Mn (n=2) ; Fe (n=2) ; Ni (n=2).

In this reaction the oxidation state of the metal cation oscillates between n and n+1.

Types

General: Irwin Fridovich and Joe McCord discovered the activity of superoxide dismutase. SODs were previously known as a group of metalloproteins with unknown function; for example, CuZnSOD was known as erythrocuprein and as the veterinary anti-inflammatory drug "Orgotein". Likewise, Brewer (1967) identified a protein that later became known as superoxide dismutase as an indophenol oxidase by protein analysis of starch gels using the phenazine-tetrazolium technique.

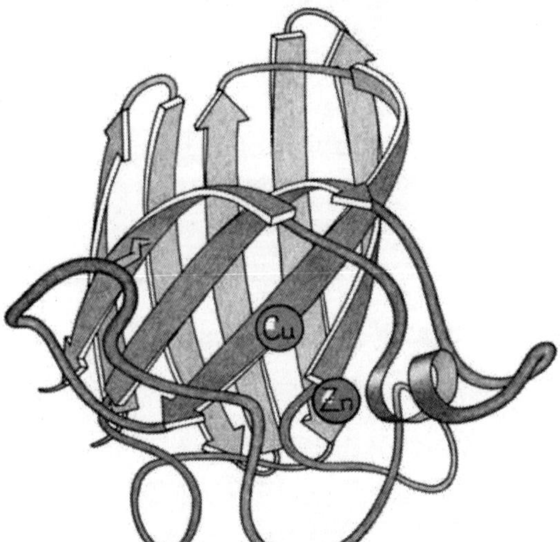

Figure: *Bovine Cu-Zn SOD subunit.*

Several common forms of SOD exist: they are proteins cofactored with copper and zinc, or manganese, iron, or nickel. Thus, there are three major families of superoxide dismutase, depending on the metal cofactor: Cu/Zn (which binds both copper and zinc), Fe and Mn types (which bind either iron or manganese), and the Ni type, which binds nickel.

- Copper and zinc – most commonly used by eukaryotes. The cytosols of virtually all eukaryotic cells contain an SOD enzyme with copper and zinc (Cu-Zn-SOD). For example, Cu-Zn-SOD available commercially is normally purified from the bovine erythrocytes: The Cu-Zn enzyme is a homodimer of molecular weight 32,500. The bovine Cu-Zn protein was the first SOD structure to be solved, in 1975. It is an 8-stranded "Greek key" beta-barrel, with the active site held between the barrel and two surface loops. The two subunits are tightly joined back-to-back, mostly by hydrophobic and some electrostatic interactions. The ligands of the copper and zinc are six histidine and one aspartate side-chains; one histidine is bound between the two metals.

- Iron or manganese – used by prokaryotes and protists, and in mitochondria
  - Iron – *E. coli* and many other bacteria also contain a form of the enzyme with iron (Fe-SOD); some bacteria contain Fe-SOD, others Mn-SOD, and some contain both. (For the *E. coli* Fe-SOD). Fe-SOD can be found in the plastids of plants. The 3D structures of the homologous Mn and Fe superoxide dismutases have the same arrangement of alpha-helices, and their active sites contain the same type and arrangement of amino acid side-chains.
  - Manganese – Chicken liver (and nearly all other) mitochondria, and many bacteria (such as *E. coli*), contain a form with manganese (Mn-SOD): For example, the Mn-SOD found in human mitochondria. The ligands of the manganese ions are 3 histidine side-chains, an aspartate side-chain and a water molecule or hydroxy ligand, depending on the Mn oxidation state (respectively II and III).
- Nickel – prokaryotic. This has a hexameric structure built from right-handed 4-helix bundles, each containing N-terminal hooks that chelate a Ni ion. The Ni-hook contains the motif His-Cys-X-X-Pro-Cys-Gly-X-Tyr; it provides most of the interactions critical for metal binding and catalysis and is, therefore, a likely diagnostic of NiSODs.

In higher plants, SOD isozymes have been localized in different cell compartments. Mn-SOD is present in mitochondria and peroxisomes. Fe-SOD has been found mainly in chloroplasts but has also been detected in peroxisomes, and CuZn-SOD has been localized in cytosol, chloroplasts, peroxisomes, and apoplast.

Plants

In higher plants, superoxide dismutase enzymes (SODs) act as antioxidants and protect cellular components from being oxidized by reactive oxygen species (ROS). ROS can form as a result of drought, injury, herbicides and pesticides, ozone, plant metabolic activity, nutrient deficiencies, photoinhibition, temperature above and below ground, toxic metals, and UV or gamma rays. To be specific, molecular O_2 is reduced to O_2^- (an ROS called superoxide) when it absorbs an excited electron released from compounds of the electron transport chain. Superoxide is known to denature enzymes, oxidize lipids, and fragment

DNA. SODs catalyze the production of O_2 and H_2O_2 from superoxide (O_2^-), which results in less harmful reactants.

When acclimating to increased levels of oxidative stress, SOD concentrations typically increase with the degree of stress conditions. The compartmentalization of different forms of SOD throughout the plant makes them counteract stress very effectively. There are three well-known and -studied classes of SOD metallic coenzymes that exist in plants. First, Fe SODs consist of two species, one homodimer (containing 1-2 g Fe) and one tetramer (containing 2-4 g Fe). They are thought to be the most ancient SOD metalloenzymes and are found within both prokaryotes and eukaryotes. Fe SODs are most abundantly localized inside plant chloroplasts, where they are indigenous. Second, Mn SODs consist of a homodimer and homotetramer species each containing a single Mn(III) atom per subunit. They are found predominantly in mitochondrion and peroxisomes. Third, Cu-Zn SODs have electrical properties very different from those of the other two classes. These are concentrated in the chloroplast, cytosol, and in some cases the extracellular space. Note that Cu-Zn SODs provide less protection than Fe SODs when localized in the chloroplast.

Bacteria

Human white blood cells generate superoxide and other reactive oxygen species to kill bacteria. During infection, some bacteria (e.g., *Burkholderia pseudomallei*) therefore produce superoxide dismutase to protect themselves from being killed.

Biochemistry

Simply stated, SOD out-competes damaging reactions of superoxide, thus protecting the cell from superoxide toxicity. The reaction of superoxide with non-radicals is spin-forbidden. In biological systems, this means that its main reactions are with itself (dismutation) or with another biological radical such as nitric oxide (NO) or with a transition-series metal. The superoxide anion radical (O_2^-) spontaneously dismutes to O_2 and hydrogen peroxide (H_2O_2) quite rapidly (~10^5 $M^{-1}s^{-1}$ at pH 7). SOD is necessary because superoxide reacts with sensitive and critical cellular targets. For example, it reacts with the NO radical, and makes toxic peroxynitrite. The dismutation rate is second-order with respect to initial superoxide concentration. Thus, the half-life of superoxide, although very short at high concentrations (e.g., 0.05 seconds at 0.1mM) is actually quite long at low concentrations (e.g., 14 hours at 0.1 nM). In contrast, the reaction of superoxide with SOD is first order with respect to superoxide concentration. Moreover, superoxide dismutase has the largest k_{cat}/K_M (an approximation of catalytic efficiency) of any known enzyme (~7 x 10^9 $M^{-1}s^{-1}$), this reaction being limited only by the frequency of collision between itself and superoxide. That is, the reaction rate is "diffusion-limited". Even at the subnanomolar concentrations achieved by the high concentrations of SOD within cells, superoxide inactivates the citric acid cycle enzyme aconitase, can poison energy metabolism, and releases potentially toxic iron. Aconitase is one of several iron-sulfur-containing (de)hydratases in metabolic pathways shown to be inactivated by superoxide.

Physiology

Superoxide is one of the main reactive oxygen species in the cell. As a consequence, SOD serves a key antioxidant role. The physiological importance of SODs is illustrated by the severe pathologies evident in mice genetically engineered to lack these enzymes. Mice lacking SOD2 die several days after birth, amid massive oxidative stress. Mice lacking SOD1 develop a wide range of pathologies, including hepatocellular carcinoma, an acceleration of age-related muscle mass loss, an earlier incidence of cataracts and a reduced lifespan. Mice lacking SOD3 do not show any obvious defects and exhibit a normal lifespan, though they are more sensitive to hyperoxic injury. Knockout mice of any SOD enzyme are more sensitive to the lethal effects of superoxide-generating drugs, such as paraquat and diquat.

Drosophila lacking SOD1 have a dramatically shortened lifespan, whereas flies lacking SOD2 die before birth. SOD knockdowns in *C. elegans* do not cause major physiological disruptions. Knockout or null mutations in SOD1 are highly detrimental to aerobic growth in the yeast Sacchormyces cerevisiae and result in a dramatic reduction in post-diauxic lifespan. SOD2 knockout or null mutations cause growth inhibition on respiratory carbon sources in addition to decreased post-diauxic lifespan.

Several prokaryotic SOD null mutants have been generated, including *E. Coli*. The loss of periplasmic CuZnSOD causes loss of virulence and might be an attractive target for new antibiotics.

Cause of Aging

According to the Free-radical theory, oxidative damage initiated by reactive oxygen species is a major contributor to the functional decline that is characteristic of aging. While studies in invertebrate models indicate that animals genetically engineered to lack specific antioxidant enzymes (such as SOD), in general, show a shortened lifespan (as one would expect from the theory), the converse manipulation, increasing the levels of antioxidant enzymes, has yielded inconsistent effects on lifespan (though some studies in Drosophila do show that lifespan can be increased by the overexpression of MnSOD or glutathione biosynthesizing enzymes). Also contrary to this theory, deletion of mitochondrial SOD2 can extend lifespan in Caenorhabditis elegans.

In mice, the story is somewhat similar. Deleting antioxidant enzymes, in general, yields shorter lifespan, though overexpression studies have not (with some recent exceptions) consistently extended lifespan.

ROS in the Context of Cancer

ROS are constantly generated and eliminated in the biological system and are required to drive regulatory pathways. Under normal physiologic conditions, cells control ROS levels by balancing the generation of ROS with their elimination by scavenging system. But under oxidative stress conditions, excessive ROS can damage cellular proteins, lipids and DNA, leading to fatal lesions in cell that contribute to carcinogenesis.

Cancer cells exhibit greater ROS stress than normal cells do, partly due to oncogenic stimulation, increased metabolic activity and mitochondrial malfunction. ROS is a double-edged sword. On one hand, at low levels, ROS facilitates cancer cell survival since cell-cycle progression driven by growth factors and receptor tyrosine kinases (RTK) require ROS for activation and chronic inflammation, a major mediator of cancer, is regulated by ROS. On the other hand, a high level of ROS can suppress tumor growth through the sustained activation of cell-cycle inhibitor and induction of cell death as well as senescence by damaging macromolecules. In fact, most of the chemotherapeutic and radiotherapeutic agents kill cancer cells by augmenting ROS stress. The ability of cancer cells to distinguish between ROS as a survival or apoptotic signal is controlled by the dosage, duration, type, and site of ROS production. Modest levels of ROS are required for cancer cells to survive, whereas excessive levels kill them.

Metabolic adaptation in tumours balances the cells' need for energy with equally important need for macromolecular building blocks and tighter control of redox balance. As a result, production of NADPH is greatly enhanced, which functions as a cofactor to provide reducing power in many enzymatic reactions for macromolecular biosynthesis and at the same time rescuing the cells from excessive ROS produced during rapid proliferation. Cells counterbalance the detrimental effects of ROS by producing antioxidant molecules, such as reduced glutathione (GSH) and thioredoxin (TRX), which rely on the reducing power of NADPH to maintain their activities.

Most risk factors associated with cancer interact with cells through the generation of ROS. ROS then activate various transcription factors such as nuclear factor kappa-light-chain-enhancer of activated B cells (NF-κB), activator protein-1 (AP-1), hypoxia-inducible factor-1α and signal transducer and activtor of transcription 3 (STAT3), leading to expression of proteins that control inflammation; cellular transformation; tumor cell survival; tumor cell proliferation; and invasion, agiogenesis as well as metastasis. And ROS also control the expression of various tumor suppressor genes such as p53, retinoblastoma gene (Rb), and phosphatase and tensin homolog (PTEN).

ROS and Carcinogenesis

ROS-related oxidation of DNA is one of the main causes of mutations, which can produce several types of DNA damage, including non-bulky (8-oxoguanine and formamidopyrimidine) and bulky (cyclopurine and etheno adducts) base modifications, abasic sites, non-conventional single-strand breaks, protein-DNA adducts, and intra/interstrand DNA crosslinks. It has been estimated that endogenous ROS produced via normal cell metabolism modify approximately 20,000 bases of DNA per day in a single cell. 8-oxoguanine is the most abundant among various oxidized nitrogeneous bases observed. During DNA replication, DNA polymerase mispairs 8-oxoguanine with adenin, leading to a G->T transition mutation. The resulting genomic instability directly contributes to carcinogenesis.

ROS and Cell Proliferation

Uncontrolled proliferation is a hallmark of cancer cells. Both exogenous and endogenous ROS have been shown to enhance proliferation of cancer cells. The role of ROS in promoting tumor proliferation is further supported by the observation that agents with potential to inhibit ROS generation can also inhibit cancer cell proliferation. Although ROS can promote tumor cell proliferation, a great increase in ROS has been associated with reduced cancer cell proliferation by induction of G2/M cell cycle arrest; increased phosphorylation of ataxia telangiectasia mutated (ATM), checkpoint kinase 1 (Chk 1), Chk 2; and reduced cell division cycle 25 homolog c (CDC25).

ROS and Cell Death

A cancer cell can die in three ways: apoptosis, necrosis and autophagy. Excessive ROS can induce apoptosis through both the extrinsic and intrinsic pathways. In the extrinsic pathway of apoptosis, ROS are generated by Fas ligand as an upstream event for Fas activation via phosphorylation, which is necessary for subsequent recruitment of

Fas-associated protein with death domain and caspase 8 as well as apoptosis induction. In the intrinsic pathway, ROS function to facilitate cytochrome c release by activating pore-stabilizing proteins (Bcl-2 and Bcl-xL) as well as inhibiting pore-destabilizing proteins (Bcl-2-associated X protein, Bcl-2 homologous antagonist/killer). An even higher ROS level can results in both apoptosis and necrosis in cancer cells. ROS can also induce cell death through autophagy, which is a self-catabolic process involving sequestration of cytoplasmic contents (exhausted organelles and protein aggregates) for degradation in lyposomes.

ROS, Tumor Cell Invasion, Angiogenesis and Metastasis

After growth factor stimulation of RTKs, ROS can trigger activation of signaling pathways involved in cell migration and invasion such as members of the mitogen activated protein kinase (MAPK) family - extracellular regulated kinase (ERK), c-jun NH-2 terminal kinase (JNK) and p38 MAPK. ROS can also promote migration by augmenting phosphorylation of the focal adhesion kinase (FAK) p130Cas and paxilin.

Both in vitro and in vivo, ROS have been shown to induce transcription factors and modulate signaling molecules involved in angiogenesis (MMP, VEGF) and metastasis (upregulation of AP-1, CXCR4, AKT and downregulation of PTEN).

ROS, Chronic Inflammation and Cancer

Experimental and epidemiologic research over the past several years has indicated close associations among ROS, chronic inflammation, and cancer. ROS induces chronic inflammation by the induction of COX-2, inflammatory cytokines (TNFα, interleukin 1 (IL-1), IL-6), chemokines (IL-8, CXCR4) and pro-inflammatory transcription factors (NF-κB). These chemokines and chemokine receptors, in turn, promote invasion and metastasis of various tumor types.

ROS and Cancer Therapy

Both ROS-elevating and ROS-eliminating strategies have been developed with the former being predominantly used. Cancer cells with elevated ROS levels depend heavily on the antioxidant defence system. ROS-elevating drugs further increase cellular ROS stress level, either by direct ROS-generation (e.g. motexafin gadolinium, elesclomol) or by agents that abrogate the inherent antioxidant system such as SOD inhibitor (e.g. ATN-224, 2-methoxyestradiol) and GSH inhibitor (e.g. PEITC, buthionine sulfoximine (BSO)). The result is an overall increase in endogenous ROS, which when above a cellular tolerability threshold, may induce cell death. On the other hand, normal cells appear to have, under lower basal stress and reserve, a higher capacity to cope with additional ROS-generating insults than cancer cells do. Therefore, preferentially accumulation ROS in cancer cells to achieve selective killing is possible.

Radiotherapy also relies on ROS toxicity to eradicate tumor cells. Radiotherapy uses X-rays, γ-rays as well as heavy particle radiation such as protons and neutrons to induce ROS-mediated cell death and mitotic failure.

Due to the dual role of ROS, both prooxidant and antioxidant-based anticancer agents have been developed. However, modulation of ROS signaling alone seems not to be an ideal approach due to adaptation of cancer cells to ROS stress, redundant pathways for supporting cancer growth and toxicity from ROS-generating anticancer drugs. Combinations of ROS-generating drugs with pharmaceuticals that can break the redox adaptation could be a better strategy for enhancing cancer cell cytotoxicity.

Antioxidant Capacity

An antioxidant is a molecule that inhibits the oxidation of other molecules. Oxidation is a chemical reaction that transfers electrons or hydrogen from a substance to an oxidizing agent. Oxidation reactions can produce free radicals. In turn, these radicals can start chain reactions. When the chain reaction occurs in a cell, it can cause damage or death to the cell. Antioxidants terminate these chain reactions by removing free radical intermediates, and inhibit other oxidation reactions. They do this by being oxidized themselves, so antioxidants are often reducing agents such as thiols, ascorbic acid, or polyphenols.

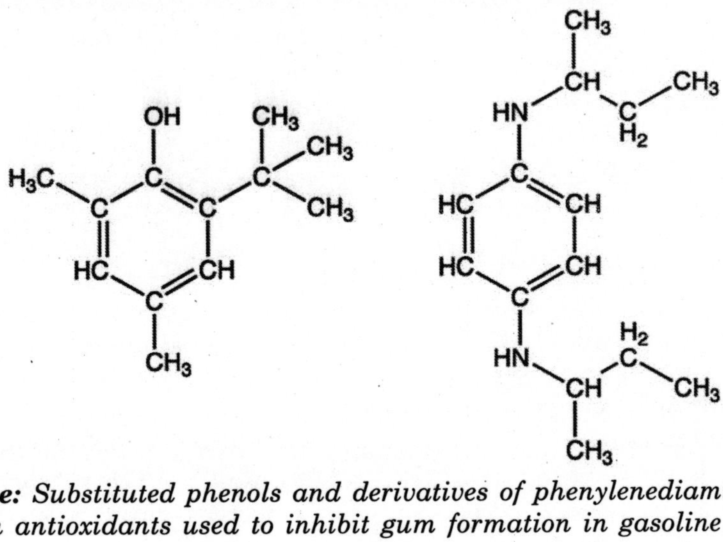

Figure: *Substituted phenols and derivatives of phenylenediamine are common antioxidants used to inhibit gum formation in gasoline (petrol).*

Although oxidation reactions are crucial for life, they can also be damaging; plants and animals maintain complex systems of multiple types of antioxidants, such as glutathione, vitamin C, vitamin A, and vitamin E as well as enzymes such as catalase, superoxide dismutase and various peroxidases. Insufficient levels of antioxidants, or inhibition of the antioxidant enzymes, cause oxidative stress and may damage or kill cells. Oxidative stress is damage to cell structure and cell function by overly reactive oxygen-containing molecules and chronic excessive inflammation. Oxidative stress seems to play a significant role in many human diseases, including cancers. The use of antioxidants in pharmacology is intensively studied, particularly as treatments for stroke and neurodegenerative diseases. For these reasons, oxidative stress can be considered to be both the cause and the consequence of some diseases.

Antioxidants are widely used in dietary supplements and have been investigated for the prevention of diseases such as cancer, coronary heart disease and even altitude sickness. Although initial studies suggested that antioxidant supplements might promote health, later large clinical trials with a limited number of antioxidants detected no benefit and even suggested that excess supplementation with certain putative antioxidants may be harmful. Antioxidants also have many industrial uses, such as preservatives in food and cosmetics and to prevent the degradation of rubber and gasoline.

Trolox Equivalent Antioxidant Capacity

Trolox equivalent antioxidant capacity (TEAC) measures the antioxidant capacity of a given substance, as compared to the standard, Trolox. Most commonly, antioxidant capacity is measured using the ABTS Decolourization Assay. Other antioxidant capacity assays which use Trolox as a standard include the diphenylpicrylhydrazyl (DPPH), oxygen radical absorbance capacity (ORAC) and ferric reducing ability of plasma (FRAP) assays. The TEAC assay is often used to measure the antioxidant capacity of foods, beverages and nutritional supplements.

Oxidative Challenge in Biology

A paradox in metabolism is that, while the vast majority of complex life on Earth requires oxygen for its existence, oxygen is a highly reactive molecule that damages living organisms by producing reactive oxygen species. Consequently, organisms contain a complex network of antioxidant metabolites and enzymes that work together to prevent oxidative damage to cellular components such as DNA, proteins and lipids. In general, antioxidant systems either prevent these reactive species from being formed, or remove them before they can damage vital components of the cell. However, reactive oxygen species also have useful cellular functions, such as redox signaling. Thus, the function of antioxidant systems is not to remove oxidants entirely, but instead to keep them at an optimum level.

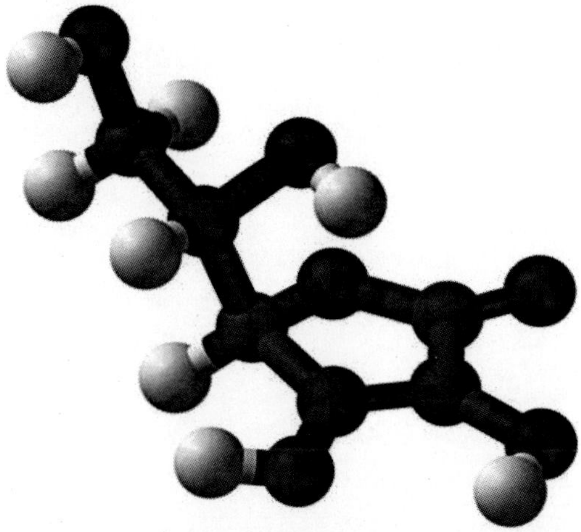

***Figure:** The structure of the antioxidant vitamin ascorbic acid (vitamin C).*

The reactive oxygen species produced in cells include hydrogen peroxide (H_2O_2), hypochlorous acid (HClO), and free radicals such as the hydroxyl radical (•OH) and the superoxide anion (O_2^{-}). The hydroxyl radical is particularly unstable and will react rapidly and non-specifically with most biological molecules. This species is produced from hydrogen peroxide in metal-catalyzed redox reactions such as the Fenton reaction. These oxidants can damage cells by starting chemical chain reactions such as lipid peroxidation, or by oxidizing DNA or proteins. Damage to DNA can cause mutations and possibly cancer, if not reversed by DNA repair mechanisms, while damage to proteins causes enzyme inhibition, denaturation and protein degradation.

The use of oxygen as part of the process for generating metabolic energy produces reactive oxygen species. In this process, the superoxide anion is produced as a by-product of several steps in the electron transport chain. Particularly important is the reduction of coenzyme Q in complex III, since a highly reactive free radical is formed as an intermediate (Q·—). This unstable intermediate can lead to electron "leakage", when electrons jump directly to oxygen and form the superoxide anion, instead of moving through the normal series of well-controlled reactions of the electron transport chain. Peroxide is also produced from the oxidation of reduced flavoproteins, such as complex I. However, although these enzymes can produce oxidants, the relative importance of the electron transfer chain to other processes that generate peroxide is unclear. In plants, algae, and cyanobacteria, reactive oxygen species are also produced during photosynthesis, particularly under conditions of high light intensity. This effect is partly offset by the involvement of carotenoids in photoinhibition, and in algae and cyanobacteria, by large amount of iodide and selenium, which involves these antioxidants reacting with over-reduced forms of the photosynthetic reaction centres to prevent the production of reactive oxygen species.

Metabolites

Overview: Antioxidants are classified into two broad divisions, depending on whether they are soluble in water (hydrophilic) or in lipids (lipophilic). In general, water-soluble antioxidants react with oxidants in the cell cytosol and the blood plasma, while lipid-soluble antioxidants protect cell membranes from lipid peroxidation.

These compounds may be synthesized in the body or obtained from the diet. The different antioxidants are present at a wide range of concentrations in body fluids and tissues, with some such as glutathione or ubiquinone mostly present within cells, while others such as uric acid are more evenly distributed. Some antioxidants are only found in a few organisms and these compounds can be important in pathogens and can be virulence factors.

The relative importance and interactions between these different antioxidants is a very complex question, with the various metabolites and enzyme systems having synergistic and interdependent effects on one another. The action of one antioxidant may therefore depend on the proper function of other members of the antioxidant system. The amount of protection provided by any one antioxidant will also depend on its concentration, its reactivity towards the particular reactive oxygen species being considered, and the status of the antioxidants with which it interacts.

Some compounds contribute to antioxidant defence by chelating transition metals and preventing them from catalyzing the production of free radicals in the cell. Particularly important is the ability to sequester iron, which is the function of iron-binding proteins such as transferrin and ferritin. Selenium and zinc are commonly referred to as *antioxidant nutrients*, but these chemical elements have no antioxidant action themselves and are instead required for the activity of some antioxidant enzymes, as is discussed below.

Antioxidant metabolite	*Solubility*	*Concentration in human serum (μM)*	*Concentration in liver tissue (μmol/kg)*
Ascorbic acid (vitamin C)	Water	50 – 60	260 (human)
Glutathione	Water	4	6,400 (human)
Lipoic acid	Water	0.1 – 0.7	4 – 5 (rat)
Uric acid	Water	200 – 400	1,600 (human)
Carotenes	Lipid	β-carotene: 0.5 – 1 retinol (vitamin A): 1 – 3	5 (human, total carotenoids)
α-Tocopherol (vitamin E)	Lipid	10 – 40	50 (human)
Ubiquinol (coenzyme Q)	Lipid	5	200 (human)

Uric Acid

Uric acid is by far the highest concentration antioxidant in human blood. Uric acid (UA) is an antioxidant oxypurine produced from xanthine by the enzyme xanthine oxidase, and is an intermediate product of purine metabolism. In almost all land animals, urate oxidase further catalyzes the oxidation of uric acid to allantoin, but in humans and most higher primates, the urate oxidase gene is nonfunctional, so that UA is not further broken down. The evolutionary reasons for this loss of urate conversion to allantoin remain the topic of active speculation. The antioxidant effects of uric acid have led researchers to suggest this mutation was beneficial to early primates and humans. Studies of high altitude acclimatization support the hypothesis that urate acts as an antioxidant by mitigating the oxidative stress caused by high-altitude hypoxia. In animal studies that investigate diseases facilitated by oxidative stress, introduction of UA both prevents the disease or reduces it, leading researchers to propose this is due to UA's antioxidant properties. Studies of UA's antioxidant mechanism support this proposal.

With respect to multiple sclerosis, Gwen Scott explains the significance of uric acid as an antioxidant by proposing that "Serum UA levels are inversely associated with the incidence of MS in humans because MS patients have low serum UA levels and individuals with hyperuricemia (gout) rarely develop the disease. Moreover, the administration of UA is therapeutic in experimental allergic encephalomyelitis (EAE), an animal model of MS." In sum, while the mechanism of UA as an antioxidant is well-supported, the claim that its levels affect MS risk is still controversial, and requires more research.

Likewise, UA has the highest concentration of any blood antioxidant and provides over half of the total antioxidant capacity of human serum. Uric acid's antioxidant activities are also complex, given that it does not react with some oxidants, such as superoxide, but does act against peroxynitrite, peroxides, and hypochlorous acid. Concerns over elevated UA's contribution to gout must be considered as one of many risk factors. By itself, UA-related risk of gout at high levels (415–530 µmol/L) is only 0.5% per year with an increase to 4.5% per year at UA supersaturation levels (535+ µmol/L). Many of these aforementioned studies determined UA's antioxidant actions within normal physiological levels, and some found antioxidant activity at levels as high as 285 µmol/L.

Ascorbic Acid (vitamin C)

Ascorbic acid or "vitamin C– is a monosaccharide oxidation-reduction (redox) catalyst found in both animals and plants. As one of the enzymes needed to make ascorbic acid has been lost by mutation during primate evolution, humans must obtain it from the diet; it is therefore a vitamin. Most other animals are able to produce this compound in their bodies and do not require it in their diets. Ascorbic acid is required for the conversion of the procollagen to collagen by oxidizing proline residues to hydroxyproline. In other cells, it is maintained in its reduced form by reaction with glutathione, which can be catalysed by protein disulfide isomerase and glutaredoxins. Ascorbic acid is a redox catalyst which can reduce, and thereby neutralize, reactive oxygen species such as hydrogen peroxide. In addition to its direct antioxidant effects, ascorbic acid is also a substrate for the redox enzyme ascorbate peroxidase, a function that is particularly important in stress resistance in plants. Ascorbic acid is present at high levels in all parts of plants and can reach concentrations of 20 millimolar in chloroplasts.

Glutathione

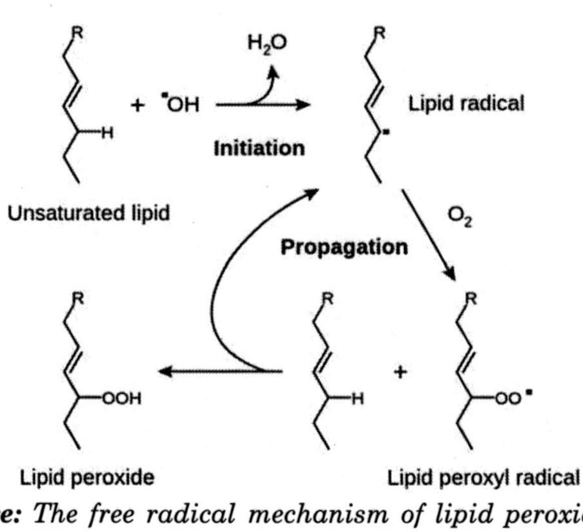

***Figure:** The free radical mechanism of lipid peroxidation.*

Glutathione is a cysteine-containing peptide found in most forms of aerobic life. It is not required in the diet and is instead synthesized in cells from its constituent amino acids. Glutathione has antioxidant properties since the thiol group in its cysteine moiety is a reducing agent and can be reversibly oxidized and reduced. In cells, glutathione is maintained in the reduced form by the enzyme glutathione reductase and in turn reduces other metabolites and enzyme systems, such as ascorbate in the glutathione-ascorbate cycle, glutathione peroxidases and glutaredoxins, as well as reacting directly with oxidants. Due to its high concentration and its central role in maintaining the cell's redox state, glutathione is one of the most important cellular antioxidants. In some organisms glutathione is replaced by other thiols, such as by mycothiol in the Actinomycetes, bacillithiol in some Gram-positive bacteria, or by trypanothione in the Kinetoplastids.

Melatonin

Melatonin is a powerful antioxidant. Melatonin easily crosses cell membranes and the blood–brain barrier. Unlike other antioxidants, melatonin does not undergo redox cycling, which is the ability of a molecule to undergo repeated reduction and oxidation. Redox cycling may allow other antioxidants (such as vitamin C) to act as pro-oxidants and promote free radical formation. Melatonin, once oxidized, cannot be reduced to its former state because it forms several stable end-products upon reacting with free radicals. Therefore, it has been referred to as a terminal (or suicidal) antioxidant.

Tocopherols and Tocotrienols (vitamin E)

Vitamin E is the collective name for a set of eight related tocopherols and tocotrienols, which are fat-soluble vitamins with antioxidant properties. Of these, α-tocopherol has been most studied as it has the highest bioavailability, with the body preferentially absorbing and metabolising this form.

It has been claimed that the α-tocopherol form is the most important lipid-soluble antioxidant, and that it protects membranes from oxidation by reacting with lipid radicals produced in the lipid peroxidation chain reaction. This removes the free radical intermediates and prevents the propagation reaction from continuing. This reaction produces oxidised α-tocopheroxyl radicals that can be recycled back to the active reduced form through reduction by other antioxidants, such as ascorbate, retinol or ubiquinol. This is in line with findings showing that α-tocopherol, but not water-soluble antioxidants, efficiently protects glutathione peroxidase 4 (GPX4)-deficient cells from cell death. GPx4 is the only known enzyme that efficiently reduces lipid-hydroperoxides within biological membranes.

However, the roles and importance of the various forms of vitamin E are presently unclear, and it has even been suggested that the most important function of α-tocopherol is as a signaling molecule, with this molecule having no significant role in antioxidant metabolism. The functions of the other forms of vitamin E are even less well-understood, although γ-tocopherol is a nucleophile that may react with electrophilic mutagens, and tocotrienols may be important in protecting neurons from damage.

Pro-oxidant Activities

Antioxidants that are reducing agents can also act as pro-oxidants. For example, vitamin C has antioxidant activity when it reduces oxidizing substances such as hydrogen peroxide, however, it will also reduce metal ions that generate free radicals through the Fenton reaction.

$$2\ Fe^{3+} + \text{Ascorbate} \rightarrow 2\ Fe^{2+} + \text{Dehydroascorbate}$$

$$2\ Fe^{2+} + 2\ H_2O_2 \rightarrow 2\ Fe^{3+} + 2\ OH\cdot + 2\ OH^-$$

The relative importance of the antioxidant and pro-oxidant activities of antioxidants are an area of current research, but vitamin C, which exerts its effects as a vitamin by oxidizing polypeptides, appears to have a mostly antioxidant action in the human body. However, less data is available for other dietary antioxidants, such as vitamin E, or the polyphenols. Likewise, the pathogenesis of diseases involving hyperuricemia likely involve uric acid's direct and indirect pro-oxidant properties.

That is, paradoxically, agents which are normally considered antioxidants can act as conditional pro-oxidants and actually increase oxidative stress. Besides ascorbate, medically important conditional pro-oxidants include uric acid and sulfhydryl amino acids such as homocysteine. Typically, this involves some transition-series metal such as copper or iron as catalyst. The potential role of the pro-oxidant role of uric acid in (e.g.) atherosclerosis and ischemic stroke is considered above. Another example is the postulated role of homocysteine in atherosclerosis.

Potential of Antioxidant Supplements to Damage Health

Some antioxidant supplements may promote disease and increase mortality in humans. Hypothetically, free radicals induce an endogenous response which protects against exogenous radicals (and possibly other toxic compounds). Recent experimental evidence strongly suggests that this is indeed the case, and that such induction of endogenous free radical production extends the life span of *Caenorhabditis elegans*. Most importantly, this induction of life span is prevented by antioxidants, providing direct evidence that toxic radicals may mitohormetically exert life extending and health promoting effects.

Enzyme Systems

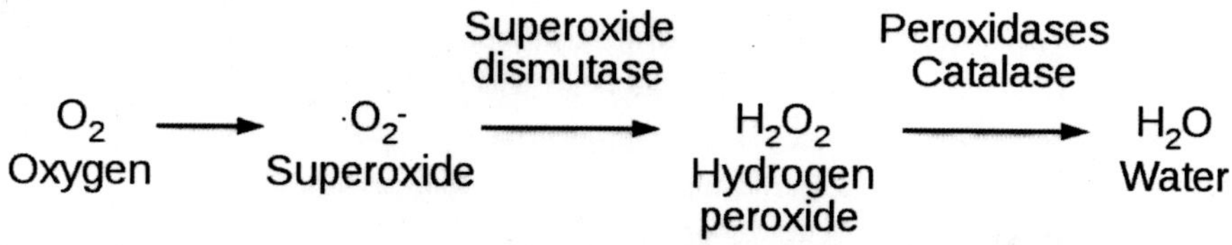

Figure: *Enzymatic pathway for detoxification of reactive oxygen species.*

As with the chemical antioxidants, cells are protected against oxidative stress by an interacting network of antioxidant enzymes. Here, the superoxide released by processes such as oxidative phosphorylation is first converted to hydrogen peroxide and then further reduced to give water. This detoxification pathway is the result of multiple enzymes, with superoxide dismutases catalysing the first step and then catalases and various peroxidases removing hydrogen peroxide. As with antioxidant metabolites, the contributions of these enzymes to antioxidant defences can be hard to separate from one another, but the generation of transgenic mice lacking just one antioxidant enzyme can be informative.

Superoxide Dismutase, Catalase and Peroxiredoxins

Superoxide dismutases (SODs) are a class of closely related enzymes that catalyze the breakdown of the superoxide anion into oxygen and hydrogen peroxide. SOD enzymes are present in almost all aerobic cells and in extracellular fluids. Superoxide dismutase enzymes contain metal ion cofactors that, depending on the isozyme, can be copper, zinc, manganese or iron. In humans, the copper/zinc SOD is present in the cytosol, while manganese SOD is present in the mitochondrion. There also exists a third form of SOD in extracellular fluids, which contains copper and zinc in its active sites. The mitochondrial isozyme seems to be the most biologically important of these three, since mice lacking this enzyme die soon after birth. In contrast, the mice lacking copper/ zinc SOD (Sod1) are viable but have numerous pathologies and a reduced lifespan, while mice without the extracellular SOD have minimal defects (sensitive to hyperoxia). In plants, SOD isozymes are present in the cytosol and mitochondria, with an iron SOD found in chloroplasts that is absent from vertebrates and yeast.

Catalases are enzymes that catalyse the conversion of hydrogen peroxide to water and oxygen, using either an iron or manganese cofactor. This protein is localized to peroxisomes in most eukaryotic cells. Catalase is an unusual enzyme since, although hydrogen peroxide is its only substrate, it follows a ping-pong mechanism. Here, its cofactor is oxidised by one molecule of hydrogen peroxide and then regenerated by transferring the bound oxygen to a second molecule of substrate. Despite its apparent importance in hydrogen peroxide removal, humans with genetic deficiency of catalase — "acatalasemia" — or mice genetically engineered to lack catalase completely, suffer few ill effects.

Figure: *Decameric structure of AhpC, a bacterial 2-cysteine peroxiredoxin from* Salmonella typhimurium.

Peroxiredoxins are peroxidases that catalyze the reduction of hydrogen peroxide, organic hydroperoxides, as well as peroxynitrite. They are divided into three classes: typical 2-cysteine peroxiredoxins; atypical 2-cysteine peroxiredoxins; and 1-cysteine peroxiredoxins. These enzymes share the same basic catalytic mechanism, in which a redox-active cysteine (the peroxidatic cysteine) in the active site is oxidized to a sulfenic acid by the peroxide substrate. Over-oxidation of this cysteine residue in peroxiredoxins inactivates these enzymes, but this can be reversed by the action of sulfiredoxin. Peroxiredoxins seem to be important in antioxidant metabolism, as mice lacking peroxiredoxin 1 or 2 have shortened lifespan and suffer from hemolytic anaemia, while plants use peroxiredoxins to remove hydrogen peroxide generated in chloroplasts.

Thioredoxin and Glutathione Systems

The thioredoxin system contains the 12-kDa protein thioredoxin and its companion thioredoxin reductase. Proteins related to thioredoxin are present in all sequenced organisms. Plants, such as *Arabidopsis thaliana,* have a particularly great diversity of isoforms. The active site of thioredoxin consists of two neighbouring cysteines, as part of a highly conserved CXXC motif, that can cycle between an active dithiol form (reduced) and an oxidized disulfide form. In its active state, thioredoxin acts as an efficient reducing agent, scavenging reactive oxygen species and maintaining other proteins in their reduced state. After being oxidized, the active thioredoxin is regenerated by the action of thioredoxin reductase, using NADPH as an electron donor.

The glutathione system includes glutathione, glutathione reductase, glutathione peroxidases and glutathione *S*-transferases. This system is found in animals, plants and microorganisms. Glutathione peroxidase is an enzyme containing four selenium-cofactors that catalyzes the breakdown of hydrogen peroxide and organic hydroperoxides. There are at least four different glutathione peroxidase isozymes in animals. Glutathione peroxidase 1 is the most abundant and is a very efficient scavenger of hydrogen peroxide, while glutathione peroxidase 4 is most active with lipid hydroperoxides. Surprisingly, glutathione peroxidase 1 is dispensable, as mice lacking this enzyme have normal lifespans, but they are hypersensitive to induced oxidative stress. In addition, the glutathione *S*-transferases show high activity with lipid peroxides. These enzymes are at particularly high levels in the liver and also serve in detoxification metabolism.

Oxidative Stress in Disease

Oxidative stress is thought to contribute to the development of a wide range of diseases including Alzheimer's disease, Parkinson's disease, the pathologies caused by diabetes, rheumatoid arthritis, and neurodegeneration in motor neuron diseases. In many of these cases, it is unclear if oxidants trigger the disease, or if they are produced as a secondary consequence of the disease and from general tissue damage; One case in which this link is particularly well-understood is the role of oxidative stress in cardiovascular disease. Here, low density lipoprotein (LDL) oxidation appears to trigger the process of atherogenesis, which results in atherosclerosis, and finally cardiovascular disease.

Oxidative damage in DNA can cause cancer. Several antioxidant enzymes such as superoxide dismutase, catalase, glutathione peroxidase, glutathione reductase, glutathione S-transferase etc. protect DNA from oxidative stress. It has been proposed that polymorphisms in these enzymes are associated with DNA damage and subsequently the individual's risk of cancer susceptibility.

A low calorie diet extends median and maximum lifespan in many animals. This effect may involve a reduction in oxidative stress. While there is some evidence to support the role of oxidative stress in aging in model organisms such as *Drosophila melanogaster* and *Caenorhabditis elegans*, the evidence in mammals is less clear. Indeed, a 2009 review of experiments in mice concluded that almost all manipulations of antioxidant systems had no effect on aging.

Diets high in fruit and vegetables, which are high in antioxidants, promote health and reduce the effects of aging; antioxidant vitamin supplementation has no detectable effect on the aging process, so the effects of fruit and vegetables may be unrelated to their antioxidant contents. One reason for this might be the fact that consuming antioxidant molecules such as polyphenols and vitamin E will produce changes in other parts of metabolism, and it may be these other effects that are the real reason these compounds are important in human nutrition.

Measurement and Levels in Food

Measurement of antioxidants is not a straightforward process, as this is a diverse group of compounds with different reactivities to different reactive oxygen species. In food science, the oxygen radical absorbance capacity (ORAC) used to be the industry standard for antioxidant strength of whole foods, juices and food additives. However, the United States Department of Agriculture (USDA) withdrew these ratings in 2012 as biologically invalid, stating that no physiological proof *in vivo* existed to support the free-radical theory. Consequently, the ORAC method, derived only in *in vitro* experiments, is no longer considered relevant to human diets or biology.

Alternative *in vitro* measurements include the Folin-Ciocalteu reagent, and the Trolox equivalent antioxidant capacity assay.

Antioxidants are found in vegetables, fruits, grain cereals, eggs, meat, legumes and nuts. Some, such as lycopene and ascorbic acid, can be destroyed by long-term storage or prolonged cooking. Other antioxidant compounds are more stable, such as the polyphenolic antioxidants in foods such as whole-wheat cereals and tea. The effects of cooking and food processing are complex, as these processes can also increase the bioavailability of antioxidants, such as some carotenoids in vegetables. Processed food contains fewer antioxidants than fresh and uncooked foods, as preparation exposes food to oxygen.

Antioxidant compounds	*Foods containing high levels of these antioxidants*
Vitamin C (ascorbic acid)	Fresh Fruits and vegetables
Vitamin E (tocopherols, tocotrienols)	Vegetable oils
Polyphenolic antioxidants (resveratrol, flavonoids)	Tea, coffee, soy, fruit, olive oil, chocolate, cinnamon, oregano
Carotenoids (lycopene, carotenes, lutein)	Fruit, vegetables and eggs.

Other antioxidants are not vitamins and are instead made in the body. For example, ubiquinol (coenzyme Q) is poorly absorbed from the gut and is made in humans through the mevalonate pathway. Another example is glutathione, which is made from amino acids. As any glutathione in the gut is broken down to free cysteine, glycine and glutamic acid before being absorbed, even large oral doses have little effect on the concentration of glutathione in the body. Although large amounts of sulfur-containing amino acids such as acetylcysteine can increase glutathione, no evidence exists that eating high levels of these glutathione precursors is beneficial for healthy adults. Supplying more of these precursors may be useful as part of the treatment of some diseases, such as acute respiratory distress syndrome, protein-energy malnutrition, or preventing the liver damage produced by paracetamol overdose.

Other compounds in the diet can alter the levels of antioxidants by acting as pro-oxidants. Here, consuming the compound causes oxidative stress, which the body responds to by inducing higher levels of antioxidant defences such as antioxidant enzymes. Some of these compounds, such as isothiocyanates and curcumin, may be chemopreventive agents that either block the transformation of abnormal cells into cancerous cells, or even kill existing cancer cells.

Polyphenol Antioxidants

A polyphenol antioxidant is a type of antioxidant containing a polyphenolic or natural phenol substructure. Numbering over 4,000 distinct species, many of these compounds have antioxidant activity in vitro but are unlikely to have antioxidant roles in vivo. Rather, they may affect cell-to-cell signaling, receptor sensitivity, inflammatory enzyme activity or gene regulation.

Evolutionary Perspective

From about 500 million years ago, freshwater and terrestrial plants slowly optimized the production of "new" endogenous antioxidants, such as ascorbic acid (vitamin C), natural phenols (including flavonoids) and polyphenols, tocopherols, etc. A few of these appeared more recently, in the last 50–200 million years, in fruits and flowers of angiosperm plants. In fact, the angiosperms (the flowering plants), the dominant type of plant today (and most of their antioxidant pigments) evolved during the late Jurassic period.

Sources of Polyphenol Antioxidants

The main source of polyphenols is dietary, since they are found in a wide array of phytochemical-bearing foods. For example, honey; most legumes; fruits such as apples, blackberries, blueberries, cantaloupe, pomegranate, cherries, cranberries, grapes, pears, plums, raspberries, aronia berries, and strawberries; and vegetables such as broccoli, cabbage, celery, onion and parsley are rich in polyphenols. Red wine, chocolate, black tea, white tea, green tea, olive oil, argan oil, bee pollen and many grains are sources. Ingestion of polyphenols occurs by consuming a wide array of plant foods.

Biochemical Theory

The regulation theory considers a polyphenol antioxidant's ability to scavenge free radicals and up-regulate certain metal chelation reactions. Various reactive oxygen species, such as singlet oxygen, peroxynitrite and hydrogen peroxide, must be continually removed from cells to maintain healthy metabolic function. Diminishing the concentrations of reactive oxygen species can have several benefits possibly associated with ion transport systems and so may affect redox signaling.

The "deactivation" of oxidant species by polyphenolic antioxidants (POH) is based, with regard to food systems that are deteriorated by peroxyl radicals (R•), on the donation of hydrogen, which interrupts chain reactions:

$$R\bullet + POH \rightarrow R\text{-}H + PO\bullet$$

Phenoxyl radicals (PO•) generated according to this reaction may be stabilized through resonance and/or intramolecular hydrogen bonding, as proposed for quercetin, or combine to yield dimerisation products, thus terminating the chain reaction:

$$PO\bullet + PO\bullet \rightarrow PO\text{-}OP$$

Potential Biological Consequences

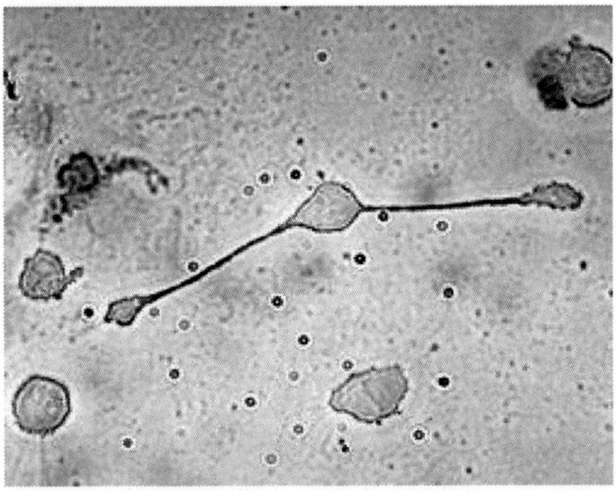

Figure: *A macrophage stretching its arms to engulf two particles. Reactive oxygen species promote oxidized LDL and polyphenol antioxidants combat this inflammatory response.*

Consuming dietary polyphenols may be associated with effects in higher animal species:

- Possible reduction in inflammation such as in coronary artery disease including specific research on endothelial cells via downregulation of oxidative LDL.
- Other possible effects may result from consumption of foods rich in polyphenols, but are not yet proved scientifically in humans so are not allowed as health statements by regulatory authorities like the U.S. Food and Drug Administration (FDA). Among these are possible anti-aging effects in skin. Further research may discern if polyphenol antioxidants have biological roles in vivo.

Difficulty in Analyzing Effects of Specific Chemicals

It is difficult to evaluate the physiological effects of specific natural phenolic antioxidants, since such a large number of individual compounds may occur even in a single food and their fate in vivo cannot be measured. For example, over sixty different chemically distinct flavonoids are known to occur in a given red wine. The polyphenol content of wines is usually evaluated by the Folin-Ciocalteu reagent which correlates well with alternative chemical and biological procedures for determining antioxidant potential.

Other more detailed chemical research has elucidated the difficulty of isolating individual phenolic antioxidants. Because significant variation in phenolic content occurs among various brands of tea, there are possible inconsistencies among epidemiological studies implying beneficial health effects of phenolic antioxidants of green tea blends. The Oxygen Radical Absorbance Capacity (ORAC) test is a laboratory indicator of antioxidant potential in foods and dietary supplements. However, ORAC results cannot be confirmed to be physiologically applicable.

Practical Aspects of Dietary Polyphenol Antioxidants

Figure: *Cocoa is the prime ingredient of chocolate, a source of polyphenols.*

There is debate regarding the total body absorption of dietary intake of polyphenolic compounds. While some indicate potential health effects of certain specific polyphenols, most studies demonstrate low bioavailability and rapid excretion of polyphenols, indicating their potential roles only in small concentrations in vivo. More research is needed to understand the interactions between a variety of these chemicals acting in concert within the human body. In particular there is evidence that some combinations of foods may inhibit efficient intestinal transfer of certain polyphenol antioxidants; refined sugars, for example, have been shown to impede this uptake under certain circumstances. Furthermore caution should be exercised in attempting diets depending largely on dietary supplements as opposed to a broad array of food sources, since the quality and concentrations of beneficial chemicals in some commercial products is subject to question.

Topical Application of Polyphenol Antioxidants

There is little evidence that reactive oxygen species play a role in the process of skin aging. The skin is exposed to various exogenous sources of oxidative stress, including ultraviolet radiation whose spectral components may be responsible for the extrinsic type of skin aging, sometimes termed photoaging. It has been shown not only that increased levels of protective low molecular weight antioxidants through a diet rich in phytochemicals, but also by direct topical dermal application of low molecular weight antioxidants, notably vitamins C and E, as well as lipoic acid, may confer protective effects against oxidative stress. However, controlled long-term studies on the efficacy of low molecular weight antioxidants in the prevention or treatment of skin aging in humans are absent.

Combination of Antioxidants in Vitro

Synergistic or Antagonistic Effects of Phenolic Mixtures: Experiments on linoleic acid subjected to 2,22 -azobis (2-amidinopropane) dihydrochloride-induced oxidation with different combinations of phenolics show that binary mixtures can lead to either a synergetic effect or to an antagonistic effect.

Antioxidant levels of purified anthocyanin extracts were much higher than expected from anthocyanin content indicating synergistic effect of anthocyanin mixtures.

While most antioxidants go directly from an active to an inactive role, emblicanins utilize a multilevel cascade of antioxidant compounds resulting in a prolongation of its antioxidant capabilities. Emblicanin A (one of the key compounds in emblicanins) aggressively reacts with free radicals. After it neutralizes a free radical, emblicanin A is transformed into emblicanin B, another antioxidant. Emblicanin B in turn also reacts with free radicals and is transformed into emblicanin oligomers. This makes emblicanins one of the best free radical scavenging antioxidant.

Effect in Combination with another Non-phenolic Antioxidant

Experiments with binary mixtures of natural phenolic antioxidants with a non-phenolic antioxidant -ascorbic acid- show that phenolics exhibit antagonism when combined with this latter compound. This is described as to the regenerating ability of the most efficient antioxidant at the expense of the less efficient one.

Nevertheless, experiments led with hydroxytyrosol, a major constituent of olive oil and α-tocopherol show a synergistic effect.

Ferricyanide

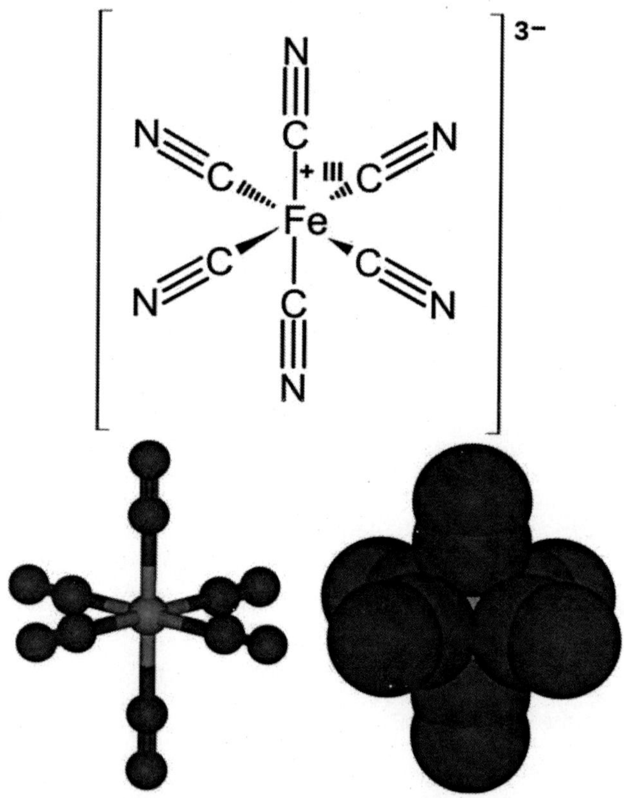

Figure: *Ferricyanide is the anion $[Fe(CN)_6]$–6. It is also called hexacyanoferrate(III) and in rare, but systematic nomenclature, hexacyanidoferrate(III). The most common salt of this anion is potassium ferricyanide, a red crystalline material that is used as an oxidant in organic chemistry.*

Properties

$[Fe(CN)_6]^{3-}$ consists of a Fe^{3+} centre bound in octahedral geometry to six cyanide ligands. The complex has O_h symmetry. The iron is low spin and easily reduced to the related ferrocyanide ion $[Fe(CN)_6]^{4-}$, which is a ferrous (Fe^{2+}) derivative. This redox couple is reversible and entails no making or breaking of Fe-C bonds:

$$[Fe(CN)_6]^{3-} + e^- \rightarrow [Fe(CN)_6]^{4-}$$

This redox couple is a standard in electrochemistry.

Compared to normal cyanides like potassium cyanide, ferricyanides are much less toxic because of the tight hold of the CN^- to the Fe^{3+}. They do react with mineral acids, however, to release highly toxic hydrogen cyanide gas.

Uses

Treatment of ferricyanide with ferrous salts affords the brilliant, long-lasting pigment Prussian blue, the traditional colour of blueprints.

DPPH

DPPH is a common abbreviation for an organic chemical compound 2,2-diphenyl-1-picrylhydrazyl. It is a dark-coloured crystalline powder composed of stable free-radical molecules.

DPPH has two major applications, both in laboratory research: one is a monitor of chemical reactions involving radicals, most notably it is a common antioxidant assay, and another is a standard of the position and intensity of electron paramagnetic resonance signals.

Properties and Applications

DPPH has several crystalline forms which differ by the lattice symmetry and melting point (m.p.). The commercial powder is a mixture of phases which melts at ~130 °C. DPPH-I (m.p. 106 °C) is orthorhombic, DPPH-II (m.p. 137 °C) is amorphous and DPPH-III (m.p. 128–129 °C) is triclinic.

DPPH is a well-known radical and a trap ("scavenger") for other radicals. Therefore, rate reduction of a chemical reaction upon addition of DPPH is used as an indicator of the radical nature of that reaction. Because of a strong absorption band centred at about 520 nm, the DPPH radical has a deep violet colour in solution, and it becomes colourless or pale yellow when neutralized.

This property allows visual monitoring of the reaction, and the number of initial radicals can be counted from the change in the optical absorption at 520 nm or in the EPR signal of the DPPH.

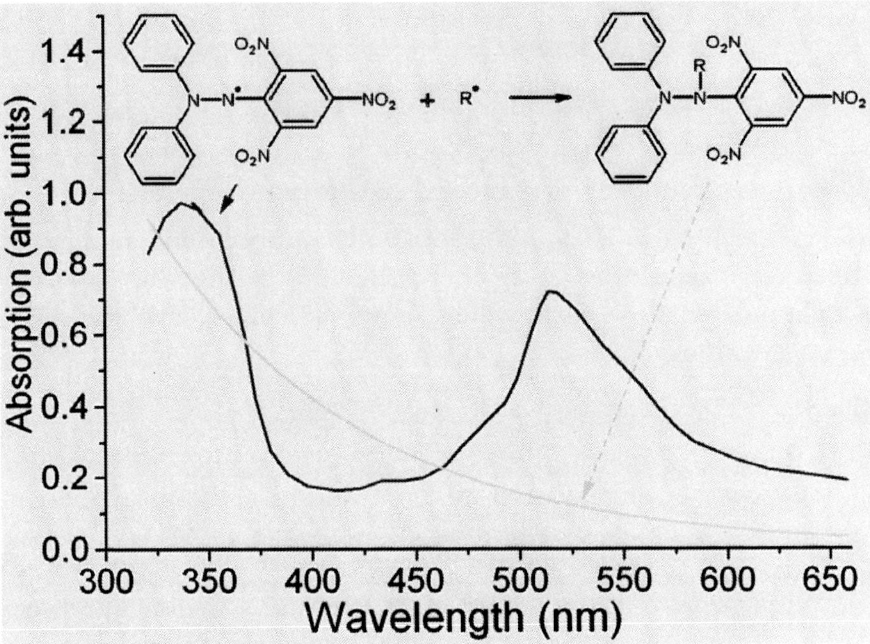

***Figure:** Because DPPH is efficient radical trap, it is also a strong inhibitor of radical-mediated polymerization.*

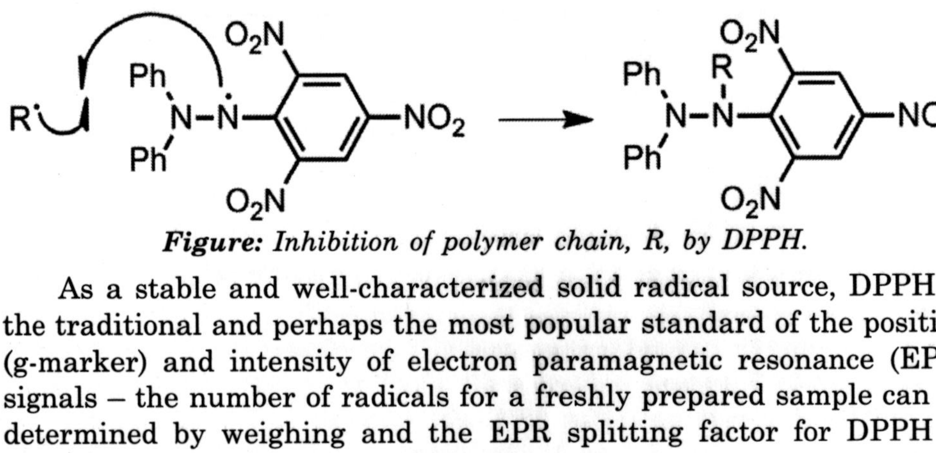

***Figure:** Inhibition of polymer chain, R, by DPPH.*

As a stable and well-characterized solid radical source, DPPH is the traditional and perhaps the most popular standard of the position (g-marker) and intensity of electron paramagnetic resonance (EPR) signals – the number of radicals for a freshly prepared sample can be determined by weighing and the EPR splitting factor for DPPH is calibrated at g = 2.0036. DPPH signal is convenient by that it is normally concentrated in a single line, whose intensity increases linearly with the square root of microwave power in the wider power range. The dilute nature of the DPPH radicals (one unpaired spin per 41 atoms) results in a relatively small linedeprecated (1.5–4.7 Gauss). The linedeprecated may however increase if solvent molecules remain in the crystal and if measurements are performed with a high-frequency EPR setup (~200 GHz), where the slight g-anisotropy of DPPH becomes detectable.

Whereas DPPH is normally a paramagnetic solid, it transforms into an antiferromagnetic state upon cooling to very low temperatures of the order 0.3 K.

Uses in Technology

Food Preservatives

Antioxidants are used as food additives to help guard against food deterioration. Exposure to oxygen and sunlight are the two main factors in the oxidation of food, so food is preserved by keeping in the dark and sealing it in containers or even coating it in wax, as with cucumbers. However, as oxygen is also important for plant respiration, storing plant materials in anaerobic conditions produces unpleasant flavors and unappealing colours. Consequently, packaging of fresh fruits and vegetables contains an ~8% oxygen atmosphere. Antioxidants are an especially important class of preservatives as, unlike bacterial or fungal spoilage, oxidation reactions still occur relatively rapidly in frozen or refrigerated food. These preservatives include natural antioxidants such as ascorbic acid (AA, E300) and tocopherols (E306), as well as synthetic antioxidants such as propyl gallate (PG, E310), tertiary butylhydroquinone (TBHQ), butylated hydroxyanisole (BHA, E320) and butylated hydroxytoluene (BHT, E321).

The most common molecules attacked by oxidation are unsaturated fats; oxidation causes them to turn rancid. Since oxidized lipids are often discoloured and usually have unpleasant tastes such as metallic or sulfurous flavors, it is important to avoid oxidation in fat-rich foods. Thus, these foods are rarely preserved by drying; instead, they are preserved by smoking, salting or fermenting. Even less fatty foods such as fruits are sprayed with sulfurous antioxidants prior to air drying. Oxidation is often catalyzed by metals, which is why fats such as butter should never be wrapped in aluminium foil or kept in metal containers. Some fatty foods such as olive oil are partially protected from oxidation by their natural content of antioxidants, but remain sensitive to photooxidation. Antioxidant preservatives are also added to fat-based cosmetics such as lipstick and moisturizers to prevent rancidity.

Industrial Uses

Antioxidants are frequently added to industrial products. A common use is as stabilizers in fuels and lubricants to prevent oxidation, and in gasolines to prevent the polymerization that leads to the formation of engine-fouling residues. In 2007, the worldwide market for industrial antioxidants had a total volume of around 0.88 million tons. This created a revenue of circa 3.7 billion US-dollars (2.4 billion Euros).

They are widely used to prevent the oxidative degradation of polymers such as rubbers, plastics and adhesives that causes a loss of strength and flexibility in these materials. Polymers containing double bonds in their main chains, such as natural rubber and polybutadiene, are especially susceptible to oxidation and ozonolysis. They can be protected by antiozonants. Solid polymer products start to crack on exposed surfaces as the material degrades and the chains break. The mode of cracking varies between oxygen and ozone attack, the former causing a "crazy paving" effect, while ozone attack produces deeper cracks aligned at right angles to the tensile strain in the product. Oxidation and UV degradation are also frequently linked, mainly because UV radiation creates free radicals by bond breakage. The free radicals then react with oxygen to produce peroxy radicals which cause yet further damage, often in a chain reaction. Other polymers susceptible to oxidation include polypropylene and polyethylene. The former is more sensitive owing to the presence of secondary carbon atoms present in every repeat unit. Attack occurs at this point because the free radical formed is more stable than one formed on a primary carbon atom. Oxidation of polyethylene tends to occur at weak links in the chain, such as branch points in low density polyethylene.

Enzymatic Antioxidant in Saline Anthropogenic Environmental Plants

Plants under natural conditions are frequently exposed to combined stressors including drought stress and desiccation, salt stress, chilling, heat shock, heavy metals, ultraviolet, radiation, air pollutants such as ozone and SO_2, mechanical stress, nutrient deprivation, pathogen attack and high light stress. A common result of most abiotic and biotic stresses is an increased production of reactive oxygen species (ROS), which frequently result in oxidative stress. The production of ROS results from pathways such as photorespiration, from the photosynthetic apparatus and from mitochondrial respiration.

The reduction of molecular oxygen to H_2O yields to intermediates reactive oxygen species (superoxide anion ($O_2^{\cdot>}$), hydroxyl radical ($OH^{\cdot}$), hydroperoxyl radical ($HO_2^{\cdot}$), hydrogen peroxide (H_2O_2), singlet oxygen ($^1O_2^{\cdot>}$,) which are potentially toxic because they are relatively reactive compared with O_2. Reactive oxygen species may lead to the unspecific oxidations of proteins, membrane lipids or DNA injury. Malonodialdehyd (MDA) content, as a product of lipid peroxidation, has been considered as an indicator of oxidative damage in the cell membrane, resulting in disruption of metabolic function and loss of cellular integrity. Many reports confirm lipid peroxidation has been associated with damages provoked by a variety of environmental stresses, i.e. with salt stress and heavy metals.

However, there is evidence that ROS also plays important roles in the plant's defence system against pathogens and any pathogenic factors, mark certain development stages such as tracheary element formations, lignification and other cross linking process in the cell wall and act as intermediate signalling molecules to regulate the expression of genes. Thus it is very important for cells to control of level of reactive oxygen species but not eliminate them completely. Defence mechanisms against free radical-inducedoxidative stress involve: (1) preventive mechanisms,(2) repair mechanisms, (3) physical defences,and (4) antioxidant defences. The plants defendagainst these ROS by the inductionof activities of certain antioxidative enzymes suchas catalase (CAT), glutathione peroxidase (GPx), glutathione reductase (GR), andsuperoxide dismutase (SOD), which scavenge reactive oxygen species.

Superoxide dismutase catalyses the reaction of dismutation of superoxide radicals: $2O_2^{\cdot>\dagger} + 2H^+ \rightarrow H_2O_2 + O_2$. SOD is a metaloprotein with metals Cu, Zn, Mn, and Fe as co-factor, Stroiński 1999). Cu-Zn-SOD (homodimer; 31-33 kD) is present in cytosol and chloroplasts. Mn-SOD (tetramer 90kD) was found in the matrix mitochondrium, whilst Fe-SOD, commonly occurring in *Procaryota*, was detected only in some plants. On the other hand catalase catalyses the reaction: $2\ H_2O_2 \rightarrow 2H_2O + O_2$. This enzyme contains three isoenzymatic forms (CAT-l, CAT-2, and CAT-3). CAT-1 and CAT-2 occur mainly in glyoxisomes, peroxysomes, and cytosol, while CAT-3 in mitochondria and cytosol.

Another important factor decomposing H_2O_2 is peroxidase of ascorbic acid (APOX) (m.w. 28-34 kD) which contains protoheme as a prosthetic group and 4 cysteines responsible for sensitivity of the enzyme to compounds blocking a thiol group. It occurs in chloroplasts and cytosol, vacuole, and in apoplasts. APOX, in particular the chloroplast isoenzyme, becomes labile if the concentration of ascorbic acid drops significantly and it decomposes under the influence of its own radical product. Under such circumstances both monodehydroascorbate (MDHAA) and dehydroascorbate (DHAA) are easily reduced along the Halliwell-Asada path. The cell defensive mechanism also depends on small-cellular antioxidants such as glutathion (GSH), proline (Pro) or ascorbic acid (AA) which also react with radicals generated in the oxidative-stress.

The origin of oxidative stress under salinity conditions is well documented in leaves, where the inhibition of Calvin cycle results in over-reduction of oxygen and formation of superoxides. Salt stress limits also gas exchange and thereby CO_2 supply to the leaf. One consequence is the over-reduction of photosynthetic-chain electron transport. This induces the generation of ROS. Little is known about the origin of oxidative stress in roots, although salt stress-related impairment of mitochondrial function is likely to be involved. Recent studies have revealed that salt-induced oxidative stress occurs in root mitochondria in a wild salt-tolerant tomato species as indicated by increased concentrations of H_2O_2 and MDA. It is now widely accepted that superoxide radicals produced during respiration, in response to salt stress, are the main precursors of mitochondrial H_2O_2 and make an important contribution to the oxidative load experienced by the cell. Root mitochondrial ROS production is increased as a result of high salinity as constraints are imposed on electron transport through mitochondrial complexes I and II.

Calcium as macroelement and important physiological chemical element reacts with heavy metals and is required for various roles in plant cell. It is implicated in the movement of cellular organelles such as the spindle apparatus and secretory vesicles, and may play a key role in integrating plant cell metabolism. However, the homeostasis of Ca is maintained principally by the action of extrusion proteins. The cytosol is strongly buffered against high concentrations of calcium by numerous ranges of calcium-binding proteins such as calmodulins and calmodulin-binding proteins. Reports showed close interaction between intracellular H_2O_2 and cytosolic calcium in response to biotic and abiotic stresses -increase in cytosolic calcium boosts the generation of H_2O_2. The protein calmodulin binds to activates plant catalases in the present of calcium. It indicated a dual function of Ca in regulating H_2O_2 homeostasis.

Metals are involved in the direct or indirect generation of free radicals (FR) and reactive oxygen species in the following ways:

1. direct transfer of electron in the single electron reduction;
2. disturbance of metabolic pathways resulting in an increase in the rate of FR and ROS formation;
3. inactivation and down regulation of the enzymes of the antioxidative defence system, and
4. depletion of low molecular weight of antioxidants.

Microelements such as Fe, Zn, Cu and Mn fulfil various roles in the metabolism of plant organism and are necessary for the regularity of physiological processes, however the excess and deficiency of these elements leads also to disturbance of ionic homeostasis. Fe, Mn, Cu, and Zn as transition metals have frequently unpaired electrons and they are,therefore, very good catalysts of oxygen reduction. In aqueous solutions at neutral pH, $O_2^{\cdot>}$ can generate H_2O_2, which can subsequently decompose to produce $^{\cdot}OH$ by the Haber-Weiss reaction in which Cu and Fe being involved: $O_2^{\cdot} + H_2O_2 \rightarrow O_2 + OH^- + {}^{\cdot}OH$. When iron is the transition metal in the Haber-Weiss reaction, it is called the Fenton reaction.

Toxic heavy metals Cd an Pb are considered as not essential metals for plant metabolism and stimulate formation of free radicals and reactive oxygen species. The ROS formation by toxic metals is indirectly rather than directly. The toxicity of Cd may result from its binding to sulfhydryl groups (S-H) of proteins leading to inhibition of activity and disruption of structure, due to perturbations in the nutrient balance and disturbance of cellular redox control. Due to a Cd inhibitory effect on the Calvin cycle, there is a decrease in NADPH utilization, resulting in the one-electron reduction of a large number of oxygen molecules on the reducing side of PSI

7

Adenine and Pyridine Nucleotides

Adenine is a nucleobase (a purine derivative) with a variety of roles in biochemistry including cellular respiration, in the form of both the energy-rich adenosine triphosphate (ATP) and the cofactors nicotinamide adenine dinucleotide (NAD) and flavin adenine dinucleotide (FAD), and protein synthesis, as a chemical component of DNA and RNA. The shape of adenine is complementary to either thymine in DNA or uracil in RNA.

Structure

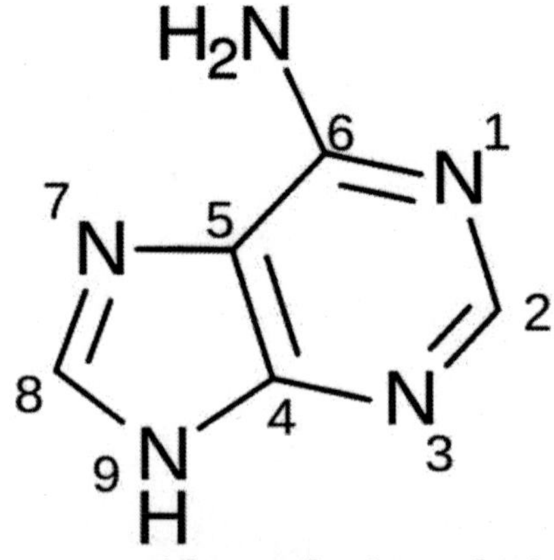

Figure: *Adenine structure, with standard numbering of positions in red.*

Adenine forms several tautomers, compounds that can be rapidly interconverted and are often considered equivalent. However, in isolated conditions, i.e. in an inert gas matrix and in the gas phase, mainly the 9H-adenine tautomer is found.

Biosynthesis

Purine metabolism involves the formation of adenine and guanine. Both adenine and guanine are derived from the nucleotide inosine monophosphate (IMP), which is synthesised on a pre-existing ribose phosphate through a complex pathway using atoms from the amino acids glycine, glutamine, and aspartic acid, as well as fused with the enzyme tetrahydrofolate.

Function

Adenine is one of the two purine nucleobases (the other being guanine) used in forming nucleotides of the nucleic acids. In DNA, adenine binds to thymine via two hydrogen bonds to assist in stabilizing the nucleic acid structures. In RNA, which is used for protein synthesis, adenine binds to uracil.

Adenine forms adenosine, a nucleoside, when attached to ribose, and deoxyadenosine when attached to deoxyribose. It forms adenosine triphosphate (ATP), a nucleotide, when three phosphate groups are added to adenosine. Adenosine triphosphate is used in cellular metabolism as one of the basic methods of transferring chemical energy between chemical reactions.

History

In older literature, adenine was sometimes called Vitamin B_4. It is no longer considered a true vitamin or part of the Vitamin B complex. However, two B vitamins, niacin and riboflavin, bind with adenine to form the essential cofactors nicotinamide adenine dinucleotide (NAD) and flavin adenine dinucleotide (FAD), respectively. Hermann Emil Fischer was one of the early scientists to study adenine.

It was named in 1885 by Albrecht Kossel, in reference to the pancreas (a specific gland - in Greek, "aden") from which Kossel's sample had been extracted.

Experiments performed in 1961 by Joan Orσ have shown that a large quantity of adenine can be synthesized from the polymerization of ammonia with five hydrogen cyanide (HCN) molecules in aqueous solution; whether this has implications for the origin of life on Earth is under debate.

On August 8, 2011, a report, based on NASA studies with meteorites found on Earth, was published suggesting building blocks of DNA and

RNA (adenine, guanine and related organic molecules) may have been formed extraterrestrially in outer space. In 2011, physicists reported that adenine has an "unexpectedly variable range of ionization energies along its reaction pathways" which suggested that "understanding experimental data on how adenine survives exposure to UV light is much more complicated than previously thought"; these findings have implications for spectroscopic measurements of heterocyclic compounds, according to one report.

Pyridine Nucleotides

Pyridine is a basic heterocyclic organic compound with the chemical formula C_5H_5N. It is structurally related to benzene, with one methine group (=CH-) replaced by a nitrogen atom. The pyridine ring occurs in many important compounds, including azines and the vitamins niacin and pyridoxal.

Pyridine was discovered in 1849 by the Scottish chemist Thomas Anderson as one of the constituents of bone oil. Two years later, Anderson isolated pure pyridine through fractional distillation of the oil. It is a colourless, highly flammable, weakly alkaline, water-soluble liquid with a distinctive, unpleasant fish-like odor.

Pyridine is used as a precursor to agrochemicals and pharmaceuticals and is also an important solvent and reagent. Pyridine is added to ethanol to make it unsuitable for drinking. It is used in the *in vitro* synthesis of DNA, in the synthesis of sulfapyridine (a drug against bacterial and viral infections), antihistaminic drugs tripelennamine and mepyramine, as well as water repellents, bactericides, and herbicides. Some chemical compounds, although not synthesized from pyridine, contain its ring structure. They include B vitamins niacin and pyridoxal, the anti-tuberculosis drug isoniazid, nicotine and other nitrogen-containing plant products. Historically, pyridine was produced from coal tar and as a by-product of the coal gasification. However, increased demand for pyridine resulted in the development of more economical methods of synthesis from acetaldehyde and ammonia, and more than 20,000 tonnes per year are manufactured worldwide.

Properties

Physical Properties: Pyridine is a colourless liquid that boils at 115.2°C and freezes at –41.6°C. Its density, 0.9819 g/cm³, is close to that of water, and its refractive index is 1.5093 at a wavelength of 589 nm and a temperature of 20°C. Addition of up to 40 mol% of water to pyridine gradually lowers its melting point from –41.6°C to –65.0°C.

The molecular electric dipole moment is 2.2 debye. Pyridine is diamagnetic and has a diamagnetic susceptibility of -48.7×10^{-6} $cm^3 \cdot mol^{-1}$.

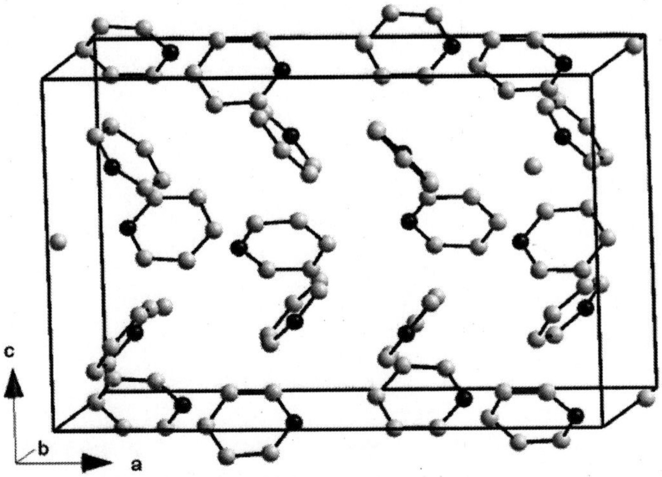

***Figure:** Crystal structure of pyridine*

The standard enthalpy of formation is 100.2 $kJ \cdot mol^{-1}$ in the liquid phase and 140.4 $kJ \cdot mol^{-1}$ in the gas phase. At 25 °C pyridine has a viscosity of 0.88 mPa/s and thermal conductivity of 0.166 $W \cdot m^{-1} \cdot K^{-1}$. The enthalpy of vaporization is 35.09 $kJ \cdot mol^{-1}$ at the boiling point and normal pressure. The enthalpy of fusion is 8.28 $kJ \cdot mol^{-1}$ at the melting point.

Pyridine crystallizes in an orthorhombic crystal system with space group $Pna2_1$ and lattice parametres a = 1752, b = 897, c = 1135 pm, and 16 formula units per unit cell (measured at 153 K). For comparison, benzene crystal is also orthorhombic, with space group *Pbca*, a = 729.2 pm, b = 947.1 pm, c = 674.2 pm (at 78 K), but the number of molecules per cell is only 4. This difference is partly related to the lower symmetry of the individual pyridine molecule (C_{2v} vs. D_{6h} for benzene). A trihydrate (pyridine $\cdot 3H_2O$) is known; it also crystallizes in an orthorhombic system in the space group *Pbca*, lattice parametres a = 1244, b = 1783, c = 679 pm and eight formula units per unit cell (measured at 223 K).

The critical parametres of pyridine are pressure 6.70 MPa, temperature 620 K and volume 229 $cm^3 \cdot mol^{-1}$. In the temperature range 340–426°C its vapor pressure p can be described with the Antoine equation

$$\log_{10} p = A - \frac{B}{C + T}$$

where T is temperature, A = 4.16272, B = 1371.358 K and C = -58.496 K.

The optical absorption spectrum of pyridine in hexane contains three bands at the wavelengths of 195 nm ($\pi \rightarrow \pi^*$transition, molar absorptivity $\varepsilon = 7500$ L·(mol·cm)$^{-1}$), 251 nm ($\pi \rightarrow \pi^*$transition, $\varepsilon = 2000$ L·(mol·cm)$^{-1}$) and 270 nm ($n \rightarrow \pi^*$transition, $\varepsilon = 450$ L·(mol·cm)$^{-1}$). The ^{1}H nuclear magnetic resonance (NMR) spectrum of pyridine contains three signals with the integral intensity ratio of 2:1:2 that correspond to the three chemically different protons in the molecule. These signals originate from the α-protons (chemical shift 8.5 ppm), γ-proton (7.5 ppm) and β-protons (7.1 ppm). The carbon analog of pyridine, benzene, has only one proton signal at 7.27 ppm. The larger chemical shifts of the α- and γ-protons in comparison to benzene result from the lower electron density in the α- and γ-positions, which can be derived from the resonance structures. The situation is rather similar for the ^{13}C NMR spectra of pyridine and benzene: pyridine shows a triplet at δ (α-C) = 150 ppm, δ (β-C) = 124 ppm and δ (γ-C) = 136 ppm, whereas benzene has a single line at 129 ppm. All shifts are quoted for the solvent-free substances. Pyridine is conventionally detected by the gas chromatography and mass spectrometry methods.

Chemical Properties

Pyridine is miscible with water and virtually all organic solvents. It is weakly basic, and with hydrochloric acid it forms a crystalline hydrochloride salt that melts at 145–147 °C. Most chemical properties of pyridine are typical of a heteroaromatic compound. In organic reactions, pyridine behaves both as a tertiary amine, undergoing protonation, alkylation, acylation, and N-oxidation at the nitrogen atom, and as an aromatic compound, undergoing nucleophilic substitutions.

Because of the electronegative nitrogen in the pyridine ring, the molecule is relatively electron deficient. It, therefore, enters less readily electrophilic aromatic substitution reactions, which are characteristic of benzene derivatives. However, unlike benzene and its derivatives, pyridine is more prone to nucleophilic substitution and metalation of the ring by strong organometallic bases. The reactivity of pyridine can be distinguished for three chemical groups. With electrophiles, electrophilic substitution takes place where pyridine expresses aromatic properties. With nucleophiles, pyridine reacts via its 2nd and 4th carbon atoms and thus behaves similar to imines and carbonyls. The reaction with many Lewis acids results in the addition to the nitrogen atom of pyridine, which is similar to the reactivity of tertiary amines. The ability of pyridine and its derivatives to oxidize, forming amine oxides (N-oxides), is also a feature of tertiary amines.

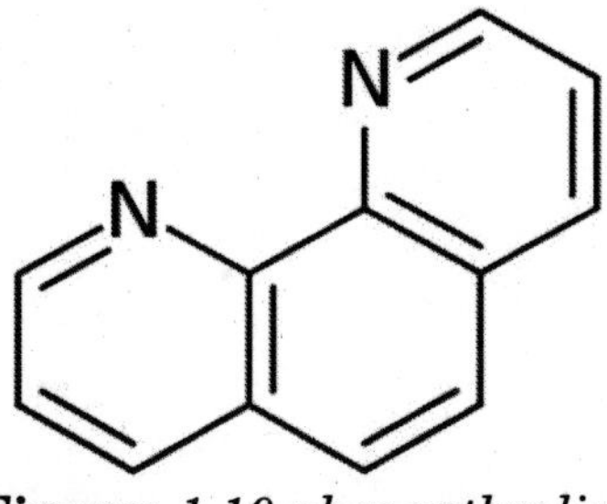

***Figure:** 1,10-phenanthroline*

The nitrogen centre of pyridine features a basic lone pair of electrons. Because this lone pair is not part of the aromatic ring, pyridine is a base, having chemical properties similar to those of tertiary amines. The pKa of the conjugate acid is 5.25. Pyridine is protonated by reaction with acids and forms a positively charged aromatic polyatomic ion called pyridinium. The bond lengths and bond angles in pyridine and pyridinium are almost identical. The pyridinium cation is isoelectronic with benzene. Pyridinium *p*-toluenesulfonate (PPTS) is an illustrative pyridinium salt; it is produced by treating pyridine with *p*-toluenesulfonic acid.

Pyridine can act as Lewis base, donating its pair of electron to a Lewis acid as in the sulfur trioxide pyridine complex.

Pyridine itself is a relatively weak ligand in forming complexes with transition metal ions. For example, it forms a 1:1 complexes with nickel(II), Ni^{2+}, and copper(II), Cu^{2+}, with $\log K_1$ values of ca. 1.9 and 2.6, respectively. The infrared spectra of pyridine complexes have been discussed in detail. Picolinic acid, which is a substituted derivative of pyridine, forms strong complexes due to the chelate effect; 2,2'-bipyridine and 1,10-phenanthroline, which can also be viewed as substituted derivatives of pyridine, also form strong complexes, such as in Ferroin, which can be used as an redox indicator in the quantitative analysis of iron.

The η^6 coordination mode, as occurs in η^6 benzene complexes, is observed only in sterically encumbered derivatives that block the nitrogen centre.

Molecular Properties

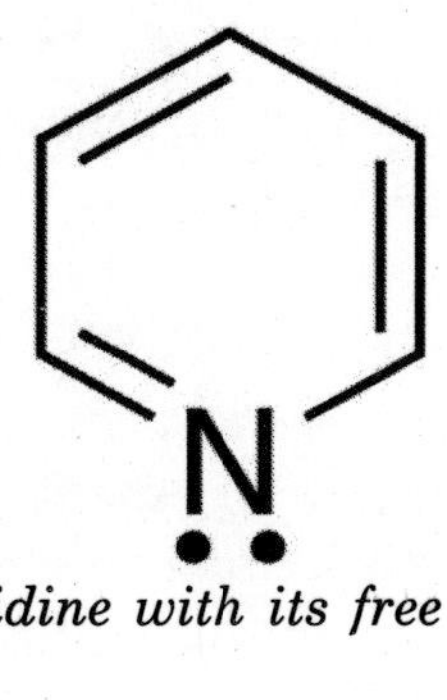

***Figure:** Pyridine with its free electron pair*

Pyridine has a conjugated system of six π-electrons that are delocalized over the ring. The molecule is planar and, thus, follows the Hückel criteria for aromatic systems. In contrast to benzene, the electron density is not evenly distributed over the ring, reflecting the negative inductive effect of the nitrogen atom. For this reason, pyridine has a dipole moment and a weaker resonant stabilization than benzene (resonance energy 117 kJ $\cdot mol^{-1}$ in pyridine vs. 150 kJ $\cdot mol^{-1}$ in benzene). The electron localization in pyridine is also reflected in the shorter C–N ring bond (137 pm for the C–N bond in pyridine vs. 139 pm for C–C bond in benzene), whereas the carbon–carbon bonds in the pyridine ring have the same 139 pm length as in benzene. These bond lengths lie between the values for the single and double bonds and are typical of aromatic compounds.

All the ring atoms in the pyridine molecule are sp^2-hybridized. The nitrogen atom "donates" its three hybridized electrons to the ring system, and its extra electron pair lies in the molecule plane, projecting outward, in the plane of the ring. This lone pair does not contribute to the aromatic system but importantly influences the chemical properties of pyridine, as it easily supports bond formation via an electrophilic attack. However, because of the separation of the lone pair from the aromatic system of the ring affects, the nitrogen atom cannot exhibit a positive mesomeric effect.

Many analogues of pyridine are known where N is replaced by other heteroatoms. Substitution of one CH in pyridine with a second N gives rise to the "diaza" heterocycles ($C_4H_4N_2$), with the names pyridazine, pyrimidine, and pyrazine.

History

Impure pyridine was undoubtedly prepared by early alchemists by heating animal bones and other organic matter, but the earliest documented reference is attributed to the Scottish scientist Thomas Anderson. In 1849, Anderson examined the contents of the oil obtained through high-temperature heating of animal bones. Among other substances, he separated from the oil a colourless liquid with unpleasant odor, from which he isolated pure pyridine two years later. He described it as highly soluble in water, readily soluble in concentrated acids and salts upon heating, and only slightly soluble in oils.

Owing to its flammability, Anderson named the new substance *pyridine*, after Greek: πæñ (pyr) meaning *fire*. The suffix *idine* was added in compliance with the chemical nomenclature, as in *toluidine*, to indicate a carbon cycle containing a nitrogen atom.

The chemical structure of pyridine was determined decades after its discovery. Wilhelm Körner (1869) and James Dewar (1871) independently suggested that, in analogy between quinoline and naphthalene, the structure of pyridine is derived from benzene by substituting one C–H unit with a nitrogen atom. The suggestion by Körner and Dewar was later confirmed in an experiment where pyridine was reduced to piperidine with sodium in ethanol. In 1876, William Ramsay combined acetylene and hydrogen cyanide into pyridine in a red-hot iron-tube furnace. This was the first synthesis of a hetero-aromatic compound.

The contemporary methods of pyridine production had a low yield, and the increasing demand for the new compound urged to search for more efficient routes. A breakthrough came in 1924 when the Russian chemist Aleksei Chichibabin invented a pyridine synthesis reaction, which was based on inexpensive reagents. This method is still used for the industrial production of pyridine.

Occurrence

Pyridine is not abundant in nature, except for the leaves and roots of belladonna (*Atropa belladonna*) and in marshmallow (*Althaea officinalis*). Pyridine derivatives, however, are often part of biomolecules such as the eponymous pyridine nucleotides and alkaloids.

In daily life, trace amounts of pyridine are components of the volatile organic compounds that are produced in roasting and canning processes, e.g. in fried chicken, sukiyaki, roasted coffee, potato chips, and fried bacon. Traces of pyridine can be found in Beaufort cheese, vaginal secretions, black tea, saliva of those suffering from gingivitis, and sunflower honey. The smoke of tobacco and marijuana also contain small amounts of pyridine.

Nomenclature

The systematic name of pyridine, within the Hantzsch–Widman nomenclature recommended by the IUPAC, is *azine*. However, systematic names for simple compounds are used very rarely; instead, heterocyclic nomenclature follows historically established common names. IUPAC discourages the use of *azine* in favour of *pyridine*. The numbering of the ring atoms in pyridine starts at the nitrogen. An allocation of positions by letter of the Greek alphabet (α-γ) and the substitution pattern nomenclature common for homoaromatic systems (*ortho*, *meta*, *para*) are used sometimes. Here α (*ortho*), β (*meta*), and γ (*para*) refer to the 2, 3, and 4 position, respectively. The systematic name for the pyridine derivatives is *pyridinyl*, wherein the position of the substituted atom is preceded by a number. However, here again the historical name *pyridyl* is encouraged by the IUPAC and used instead of the systematic name. The cationic derivative formed by the addition of an electrophile to the nitrogen atom is called *pyridinium*.

Production

Historically, pyridine was extracted from coal tar or obtained as a by-product of coal gasification. The process was labour-consuming and inefficient: coal tar contains only about 0.1% pyridine, and therefore a multi-stage purification was required, which further reduced the output. Nowadays, most pyridine is produced synthetically using various name reactions, and the major ones are discussed below.

Chichibabin Synthesis

The Chichibabin pyridine synthesis was reported in 1924 and is still in use in industry. In its general form, the reaction can be described as a condensation reaction of aldehydes, ketones, α,β-Unsaturated carbonyl compounds, or any combination of the above, in ammonia or ammonia derivatives. In particular, unsubstituted pyridine is produced from formaldehyde and acetaldehyde, which are inexpensive and widely available. First, acrolein is formed in a Knoevenagel condensation from the acetaldehyde and formaldehyde. It is then condensed with acetaldehyde and ammonia into dihydropyridine, and then oxidized with a solid-state catalyst to pyridine. This process is carried out in a gas phase at 400–450 °C. The product consists of a mixture of pyridine, simple methylated pyridines (picoline), and lutidine; its composition depends on the catalyst used and can be adapted to the needs of the manufacturer. The catalyst is usually a transition metal salt such as cadmium(II) fluoride or manganese(II) fluoride, but cobalt and thallium compounds can also be used. The recovered pyridine is separated from by-products in a multistage process.

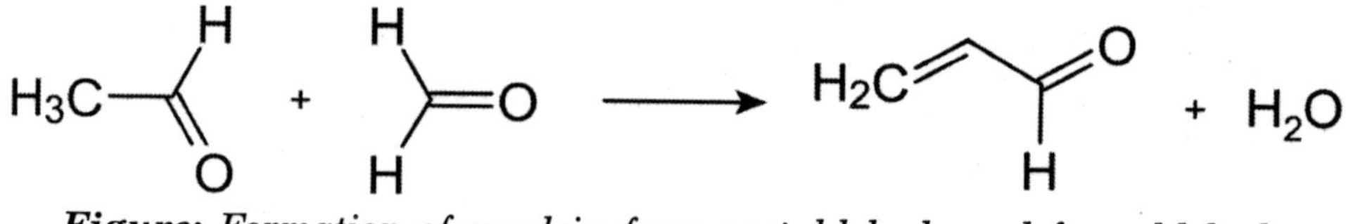

Figure: Formation of acrolein from acetaldehyde and formaldehyde

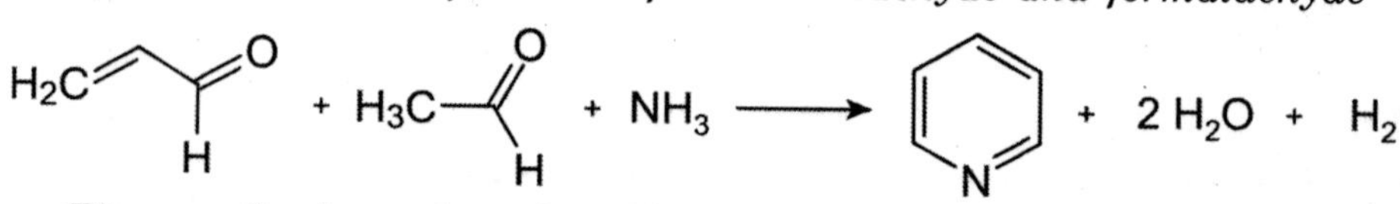

Figure: Condensation of pyridine from acrolein and acetaldehyde

Practical application of the traditional Chichibabin pyridine synthesis are limited by its consistently low yield, typically about 20%. This low yield, together with the high prevalence of byproducts, render unmodified forms of Chichibabin's method unpopular.

Dealkylation of Alkylpyridines

Pyridine can be prepared by dealkylation of alkylated pyridines, which are obtained as by-products in the syntheses of other pyridines. The oxidative dealkylation is carried out either using air over vanadium(V) oxide catalyst, by vapor-dealkylation on nickel-based catalyst, or hydrodealkylation with a silver- or platinum-based catalyst. Yields of pyridine up to be 93% can be achieved with the nickel-based catalyst.

Hantzsch Pyridine Synthesis

The first major synthesis of pyridine derivatives was described in 1881 by Arthur Rudolf Hantzsch. The Hantzsch pyridine synthesis typically uses a 2:1:1 mixture of a β-keto acid (often acetoacetate), an aldehyde (often formaldehyde), and ammonia or its salt as the nitrogen donor. First, a double hydrogenated pyridine is obtained, which is then oxidized to the corresponding pyridine derivative. Emil Knoevenagel showed that unsymmetrically substituted pyridine derivatives can be produced with this process.

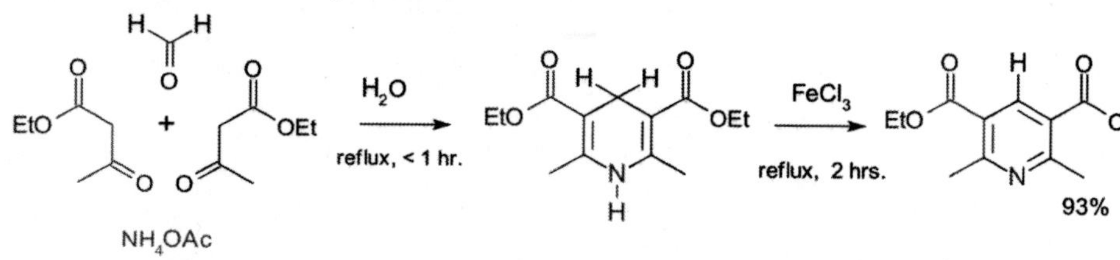

***Figure:** Hantzsch pyridine synthesis with acetoacetate, formaldehyde and ammonium acetate, and iron(III) chloride as the catalyst.*

Bönnemann Cyclization

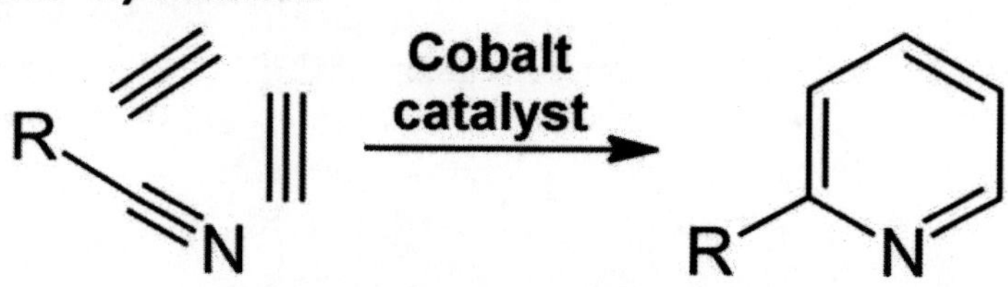

***Figure:** Bönnemann cyclization*

The trimerization of a part of a nitrile molecule and two parts of acetylene into pyridine is called Bönnemann cyclization. This modification of the Reppe synthesis can be activated either by heat or by light. While the thermal activation requires high pressures and temperatures, the photoinduced cycloaddition proceeds at ambient conditions with $CoCp_2(cod)$ (Cp = cyclopentadienyl, cod = 1,5-cyclooctadiene) as a catalyst, and can be performed even in water. A series of pyridine derivatives can be produced in this way. When using acetonitrile as the nitrile, 2-methylpyridine is obtained, which can be dealkylated to pyridine.

Other Methods

The Kröhnke pyridine synthesis involves the condensation of 1,5-diketones with ammonium acetate in acetic acid followed by oxidation.

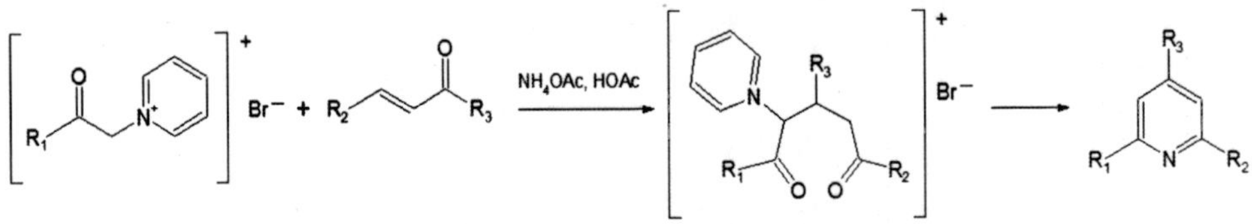

The Ciamician-Dennstedt rearrangement entails the ring-expansion of pyrrole with dichlorocarbene to 3-chloropyridine.

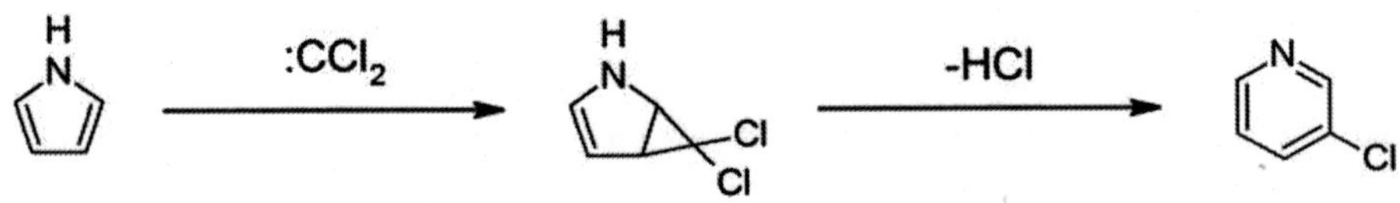

In the Gattermann-Skita synthesis, a malonate ester salt reacts with dichloromethylamine.

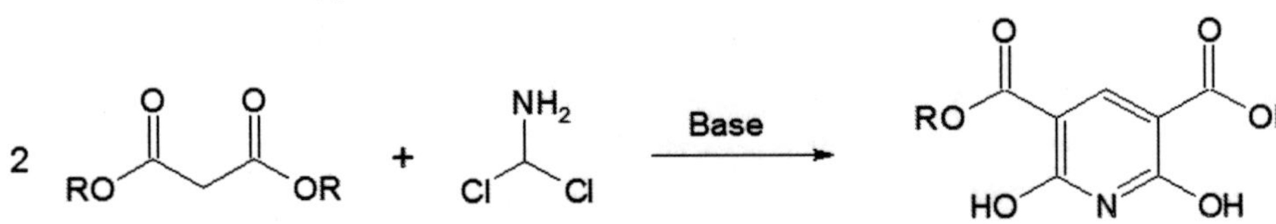

Biosynthesis

Several pyridine derivatives play important roles in biological systems. While its biosynthesis is not fully understood, nicotinic acid (vitamin B_3) occurs in some bacteria, fungi, and mammals. Mammals synthesize nicotinic acid through oxidation of the amino acid tryptophan, where an intermediate product, aniline, creates a pyridine derivative, kynurenine. On the contrary, the bacteria *Mycobacterium tuberculosis* and *Escherichia coli* produce nicotinic acid by condensation of glyceraldehyde 3-phosphate and aspartic acid.

Reactions

Many reactions that are characteristic of benzene proceed with pyridine either at more complicated conditions or/and with low yield.

Owing to the decreased electron density in the aromatic system, electrophilic substitutions are suppressed in pyridine and its derivatives in favour of addition of nucleophiles at the electron-rich nitrogen atom. The nucleophilic addition at the nitrogen atom leads to a further deactivation of the aromatic properties and hindering of the electrophilic substitution. On the other hand, free-radical and nucleophilic substitutions occur more readily in pyridine than in benzene.

Electrophilic Substitutions

Many electrophilic substitutions on pyridine either do not proceed or proceed only partially; however, the heteroaromatic character can be activated by electron-donating functionalization. Common alkylations and acylations, such as Friedel–Crafts alkylation or acylation, usually fail for pyridine because they lead only to the addition at the nitrogen atom. Substitutions usually occur at the 3-position, which is the most electron-rich carbon atom in the ring and is, therefore, more susceptible to an electrophilic addition.

Substitutions to pyridine at the 2- or 4-position result in an energetically unfavourable σ complex. They can be promoted, however, using clever experimental techniques, such as conducting electrophilic substitution on the pyridine-N-oxide followed by deoxygenation of the nitrogen atom. Addition of oxygen reduces electron density on the nitrogen atom and promotes substitution at the 2- and 4-carbons. The oxygen atom can then be removed via several routes, most commonly with compounds of trivalent phosphorus or divalent sulfur, which are easily oxidized. Triphenylphosphine is a frequently used reagent, which is oxidized in this reaction to triphenylphosphine oxide. The following paragraphs describe representative electrophilic substitution reactions of pyridine.

Direct nitration of pyridine requires harsh conditions and has very low yields. The 3-nitropyridine can be obtained instead by reacting pyridine with dinitrogen pentoxide in presence of sodium. Pyridine derivatives wherein the nitrogen atom is screened sterically and/or electronically can be obtained by nitration with nitronium tetrafluoroborate (NO_2BF_4). In this way, 3-nitropyridine can be obtained via the synthesis of 2,6-dibromopyridine followed by removal of the bromine atoms. Direct sulfonation of pyridine is even more difficult than direct nitration. However, pyridine-3-sulfonic acid can be obtained at acceptable yield by boiling pyridine in an excess of sulfuric acid at 320°C. Reaction with the SO_3 group also facilitates addition of sulfur to the nitrogen atom, especially in the presence of a mercury(II) sulfate catalyst.

In contrast to the nitration and sulfonation, the direct bromination and chlorination of pyridine proceed well. The reaction of pyridine with molecular bromine in sulfuric acid at 130°C readily produced 3-bromopyridine. The yield is lower for 3-chloropyridine upon chlorination with molecular chlorine in the presence of aluminium chloride at 100°C. Both 2-bromopyridine and 2-chloropyridine can be produced by direct reaction with halogen with a palladium(II) chloride catalyst.

Nucleophilic Substitutions

In contrast to benzene, pyridine efficiently supports several nucleophilic substitutions, and is regarded as a good nucleophile (donor number 33.1). The reason for this is relatively lower electron density of the carbon atoms of the ring. These reactions include substitutions with elimination of a hydride ion and elimination-additions with formation of an intermediate aryne configuration, and usually proceed at 2- or 4-position.

Many nucleophilic substitutions occur easier not with bare pyridine but with pyridine modified with bromine, chlorine, fluorine, or sulfonic acid fragments that then become a leaving group. So fluorine is the best leaving group for the substitution with organolithium compounds. The nucleophilic attack compounds may be alkoxides, thiolates, amines, and ammonia (at elevated pressures).

In general, the hydride ion is a poor leaving group and occurs only in a few heterocyclic reactions. They include the Chichibabin reaction, which yields pyridine derivatives aminated at the 2-position. Here, sodium amide is used as the nucleophile yielding 2-aminopyridine. The hydride ion released in this reaction combines with a proton of an available amino group, forming a hydrogen molecule.

Analogous to benzene, nucleophilic substitutions to pyridine can result in the formation of pyridyne intermediates as heteroaryne. For this purpose, pyridine derivatives can be eliminated with good leaving groups using strong bases such as sodium and potassium tert-butoxide. The subsequent addition of a nucleophile to the triple bond has low selectivity, and the result is a mixture of the two possible adducts.

Radical Reactions

Pyridine supports a series of radical reactions, which is used in its dimerization to bipyridines. Radical dimerization of pyridine with elemental sodium or Raney nickel selectively yields 4,4'-bipyridine, or 2,2'-bipyridine, which are important precursor reagents in the chemical industry. One of the name reactions involving free radicals is the

Minisci reaction. It can produce 2-*tert*-butylpyridine upon reacting pyridine with pivalic acid, silver nitrate and ammonium in sulfuric acid with a yield of 97%.

Reactions on the Nitrogen Atom

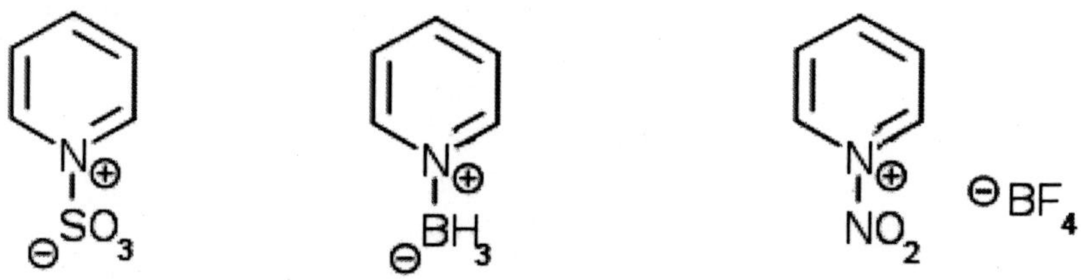

Figure: *Additions of various Lewis acids to pyridine*

Lewis acids easily add to the nitrogen atom of pyridine, forming pyridinium salts. The reaction with alkyl halides leads to alkylation of the nitrogen atom. This creates a positive charge in the ring that increases the reactivity of pyridine to both oxidation and reduction. The Zincke reaction is used for the selective introduction of radicals in pyridinium compounds (it has no relation to the chemical element zinc).

Hydrogenation and Reduction

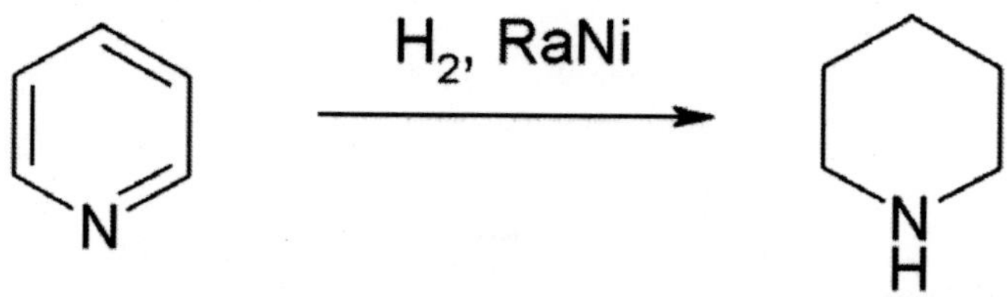

Figure: *Reduction of pyridine to piperidine with Raney nickel*

Hydrogen-saturated piperidine is obtained through reaction with hydrogen gas in the presence of Raney nickel. This reaction releases 193.8 kJ $\cdot$mol^{-1} of energy, which is slightly less than the energy of the hydrogenation of benzene (205.3 kJ $\cdot$mol^{-1}).

Partially hydrogenated derivatives are obtained under milder conditions. For example, reduction with lithium aluminium hydride yields a mixture of 1,4-dihydropyridine, 1,2-dihydropyridine, and 2,5-dihydropyridine. Selective synthesis of 1,4-dihydropyridine is achieved in the presence of organometallic complexes of magnesium and zinc, and (Δ3,4)-tetrahydropyridine is obtained by electrochemical reduction of pyridine.

Applications

Pyridine is an important raw material in of the chemical industry, with the 1989 production of 26,000 tonnes in world-wide. Among major 25 production sites for pyridine eleven are located in Europe (as of

1999). The major producers of pyridine include Evonik Industries, Rütgers Chemicals, Imperial Chemical Industries, and Koei Chemical.

The pyridine production has significantly increased in the early 2000s, with an annual production capacity of 30,000 tonnes in mainland China alone. The US-Chinese joint venture Vertellus is currently the world leader in pyridine production.

Pesticides

The main use of pyridine is as a precursor to the herbicides paraquat and diquat. The first synthesis step of insecticide chlorpyrifos consists of the chlorination of pyridine. Pyridine is also the starting compound for the preparation of pyrithione-based fungicides. Cetylpyridinium and laurylpyridinium, which can be produced from pyridine with a Zincke reaction, are used as antiseptic in oral and dental care products.Pyridine is easily attacked by alkylating agents to give *N*-alkylpyridinium salts. One example is cetylpyridinium chloride.

Solvent

Pyridine is used as polar, basic, low-reactive solvent, for example in Knoevenagel condensations. It is especially suitable for the dehalogenation, where it acts as the base of the elimination reaction and bonds the resulting hydrogen halide to form a pyridinium salt. In esterifications and acylations pyridine activates the carboxylic acid halides or anhydrides. Even more active in these reactions are the pyridine derivatives 4-dimethylaminopyridine (DMAP) and 4-(1-pyrrolidinyl) pyridine. Pyridine is also used as a base in condensation reactions.

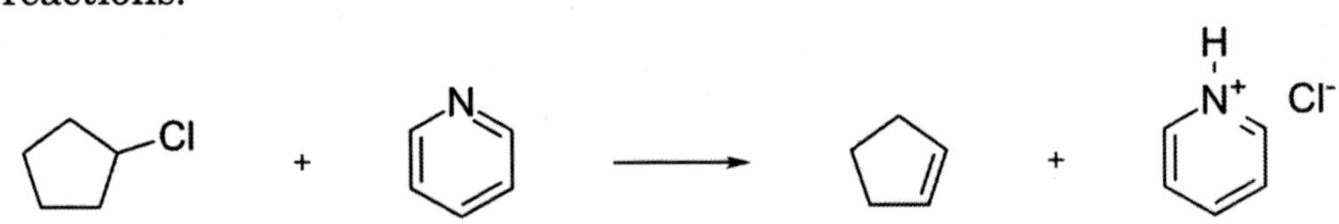

Figure: *Elimination reaction with pyridine to form pyridinium*

Pyridine is used as a solvent in the manufacture of dyes and rubber. It is also used in the textile industry to improve network capacity of cotton. Pyridine is added to ethanol to make it unsuitable for drinking. In low doses, pyridine is added to foods to give them a bitter flavour, and such usage is approved by the US Food and Drug Administration. The detection threshold for pyridine in solutions is about 1–3 mmole $\cdot L^{-1}$ (79–237 mg $\cdot L^{-1}$). As a base, pyridine can be used as the Karl Fischer reagent, but it is usually replaced by alternatives with a more pleasant odor, such as imidazole.

Precursor to Piperidine

Piperidine, a foundational nitrogen-heterocycle, is important synthetic building-block. Piperidines are produced by hydrogenation of pyridine with a nickel-, cobalt-, or ruthenium-based catalyst at elevated temperatures.

Speciality Reagents based on Pyridine

Pyridinium chlorochromate was developed by Elias James Corey and William Suggs in 1975 and is used to oxidize primary alcohols to aldehydes and secondary alcohols to ketones. It is obtained by adding pyridine to a solution of chromic acid and concentrated hydrochloric acid:

$$C_5H_5N + HCl + CrO_3 \rightarrow [C_5H_5NH][CrO_3Cl]$$

The carcinogenicity of the side-product chromyl chloride (CrO_2Cl_2) urged to look for alternative routes, such as treating chromium(VI) oxide with pyridinium chloride:

$$[C_5H_5NH^+]Cl^- + CrO_3 \rightarrow [C_5H_5NH][CrO_3Cl]$$

The Cornforth reagent (pyridinium dichromate, PDC), pyridinium chlorochromate (PCC), the Collins reagent (complex of chromium(VI) oxide with pyridine in dichloromethane) and the Sarret reagent (complex of chromium(VI) oxide with pyridine in pyridine) are similar chromium-based pyridine compounds, which are also used for oxidation, namely conversion of primary and secondary alcohols to ketones. The Collins and Sarret reagents are both difficult and dangerous to prepare, they are hygroscopic and can inflame during preparation. For this reason, the use of PCC and PDC was preferred. Those reagents were rather popular in the 1970s–1980s, but because of their toxicity and confirmed carcinogenic status, they are rarely used nowadays.

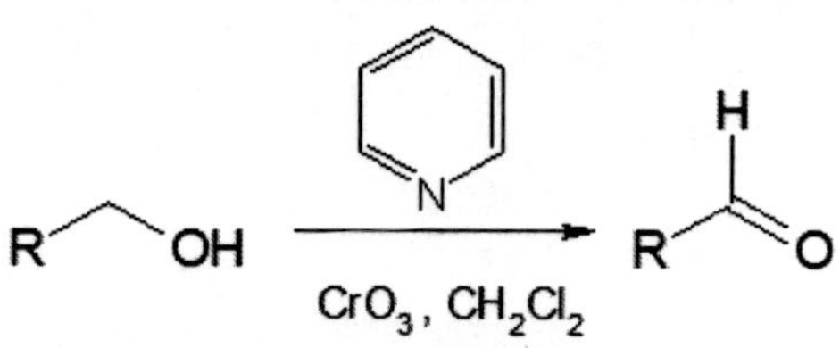

***Figure:** Oxidation of an alcohol to aldehyde with the Collins reagent.*

Ligand and Lewis Base

Pyridine is widely used as a ligand in coordination chemistry, as are its derivative 2,2'-bipyridine, consisting of two pyridine molecules joined by a single bond, and terpyridine, a molecule of three pyridine rings linked together.

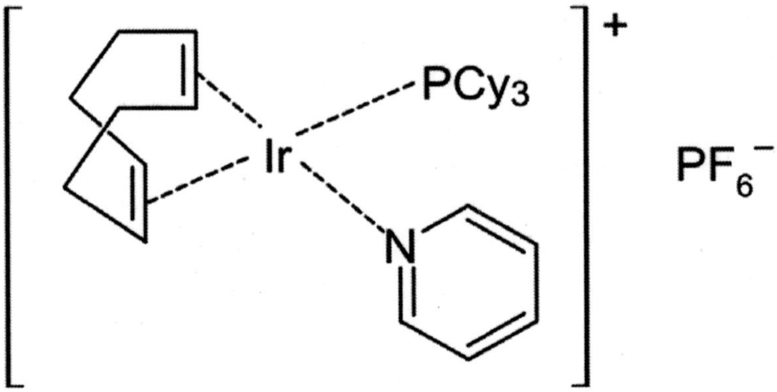

Figure: *structure of the Crabtree's catalyst*

When a pyridine ligand is part of a metal complex, it can be easily replaced by a stronger Lewis base. This property is exploited in catalysis of polymerization and hydrogenation reactions, using, for example, the Crabtree's catalyst. The pyridine ligand replaced during the reaction is restored after its completion.

Useful adducts of pyridine include pyridine-borane, $C_5H_5NBH_3$ (melting point 10–11 °C), a mild reducing agent with improved stability relative to $NaBH_4$ in protic solvents and improved solubility in aprotic organic solvents. Pyridine-sulfur trioxide, $C_5H_5NSO_3$ (melting point 175 °C) is a sulfonation agent used to convert alcohols to sulfonates, which in turn undergo C-O bond scission upon reduction with hydride agents.

Hazards

Pyridine has a flash point (the lowest temperature at which it can vaporize to form an ignitable mixture in air) of only 17 °C and is, therefore, highly flammable. Its ignition temperature is 550°C, and mixtures of 1.7–10.6 vol% of pyridine with air are explosive.

The thermal modification of pyridine starts above 490°C, resulting in bipyridine (mainly 2,2'-bipyridine and to a lesser extent 2,3'-bipyridine and 2,4'-bipyridine), nitrogen oxides, and carbon monoxide. Pyridine easily dissolves in water and harms both animals and plants in aquatic systems.

The permitted maximum allowable concentration of pyridine was 15–30 parts per million (ppm, or 15–30 mg·m^{-3} in air) in most countries in the 1990s, but was reduced to 5 ppm in the 2000s. For comparison, indoor air contaminated with tobacco smoke may contain up to 16 μg·m^{-3}, and one cigarette contains 21–32 μg of pyridine.

Health Issues

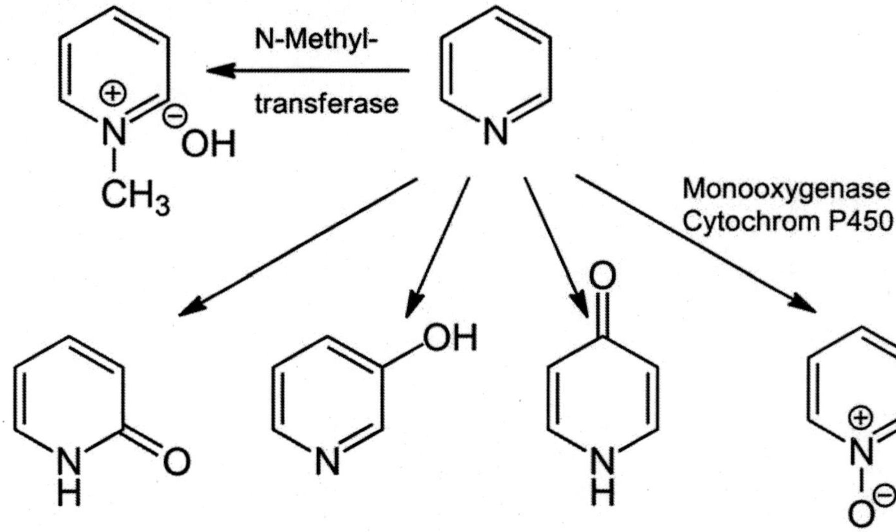

***Figure:** Metabolism of pyridine*

Pyridine is harmful if inhaled, swallowed or absorbed through the skin. Effects of acute pyridine intoxication include dizziness, headache, lack of coordination, nausea, salivation, and loss of appetite. They may progress into abdominal pain, pulmonary congestion and unconsciousness. One person died after accidental ingestion of half a cup of pyridine. The lowest known lethal dose (LD_{Lo}) for the ingestion of pyridine in humans is 500 mg·kg^{-1}. In high doses, pyridine has a narcotic effect and its vapor concentrations of above 3600 ppm pose a health risk. The oral LD_{50} in rats is 891 mg·kg^{-1}. Pyridine is flammable.

Evaluations as a possible carcinogenic agent showed that there is inadequate evidence in humans for the carcinogenicity of pyridine, although there is limited evidence of carcinogenic effects on animals. Available data indicate that "exposure to pyridine in drinking-water led to reduction of sperm motility at all dose levels in mice and increased estrous cycle length at the highest dose level in rats".

Pyridine might also have minor neurotoxic, genotoxic, and clastogenic effects. Exposure to pyridine would normally lead to its inhalation and absorption in the lungs and gastrointestinal tract, where it either remains unchanged or is metabolized. The major products of pyridine metabolism are *N*-methylpyridiniumhydroxide, which are formed by N-methyltransferases (e.g., pyridine N-methyltransferase), as well as pyridine-*N* oxide, and 2-, 3-, and 4-hydroxypyridine, which are generated by the action of monooxygenase. In humans, pyridine is metabolized only into *N*-methylpyridiniumhydroxide. Pyridine is readily degraded by bacteria to ammonia and carbon dioxide. The unsubstituted pyridine ring degrades more rapidly than picoline, lutidine, chloropyridine, or aminopyridines, and a number of pyridine degraders have been shown to overproduce riboflavin in the presence of pyridine.

Minor amounts of pyridine are released into environment from some industrial processes such as steel manufacture, processing of oil shale, coal gasification, coking plants and incinerators. The atmosphere at oil shale processing plants can contain pyridine concentrations of up to 13 $\mu g \cdot m^{-3}$, and 53 $\mu g \cdot m^{-3}$ levels were measured in the groundwater in the vicinity of a coal gasification plant.

B-Vitamins in Soil

B vitamins are a group of water-soluble vitamins that play important roles in cell metabolism. Though these vitamins share similar names, research shows that they are chemically distinct vitamins that often coexist in the same foods. In general, supplements containing all eight are referred to as a vitamin B complex. Individual B vitamin supplements are referred to by the specific name of each vitamin (e.g., B_1, B_2, B_3 etc.).

Most of the B vitamins have been shown to occur in arable soil, their presence being attributable to release from vitamin-containing plant and animal residues, liberation from the roots of growing plants and synthesis by micro-organisms. It is therefore not surprising that bacteria are present for which certain vitamins are essential for growth. It has been previously shown that among the indigenous soil bacteria are forms requiring one or more preformed vitamins, while other reports have pointed to the need of other types for certain specific factors, namely vitamin B_{12} and the terregens factor.

Studies on the production of B vitamins by soil, rhizosphere, and mycorrhizosphere bacteria have revealed that such ability is common among these organisms. Thiamin, nicotinic and folic acids were produced by the greatest number of bacteria. Bacteria isolated from the rhizosphere and mycorrhizosphere produced more riboflavin than those originating from the soil distant from roots. In general the soil strains produced higher amounts of vitamins (except riboflavin) than the root-zone isolates. Two and three vitamins were produced by the majority of bacteria. The synthesis of four and five vitamins was found in few strains only.

List of B vitamins

- Vitamin B_1 (thiamine)
- Vitamin B_2 (riboflavin)

- Vitamin B_3 (niacin or niacinamide)
- Vitamin B_5 (pantothenic acid)
- Vitamin B_6 (pyridoxine, pyridoxal, or pyridoxamine, or pyridoxine hydrochloride)
- Vitamin B_7 (biotin)
- Vitamin B_9 (folic acid)
- Vitamin B_{12} (various cobalamins; commonly cyanocobalamin in vitamin supplements)

Vitamin B 12 in Soil

B12 is an essential water soluble vitamin that plays a key role in the normal functioning of the brain, nervous system and formation of blood. Basically, not getting enough B12 can get you into some deep doo doo. However, B12 is absorbed in extremely minuscule amounts with required daily rations of less than one ounce. Also, the body holds reserves that can supply itself with enough B12 for years depending on the amount normally consumed.

B12 is found naturally in manure fertilized soil and synthesized via bacteria which are present in the intestines of animals- yummy ay? However, animals themselves are not capable of synthesizing the vitamin, only the bacteria contain the enzymes capable of breaking it down. Meaning that in order for humans to properly acquire an adequate amount of B12 they must either obtain it from animal sources, fortified foods or multi-vitamins.

While B12 is found in some plant sources including soy, it is highly debated whether or not a reliable source of B12 can be obtained directly through these plants as the traces of it are said to not be effectively absorbed directly due to a lack of the bacterial synthesization process (found in humans). Some alternative sources for Vegans include: Nutritional yeast, flax seed and other fortified foods or multi-vitamins. Basically, animal eats plant and absorbs and breaks down B12 via bacteria which is found in the soil, fertilized by animal manure and (some) people eat animal and absorb already synthesized B12.

Often the above conclusions lead some to a single one: That eating animal products is “natural” to the human diet. However, the evidence surrounding this statement is equally as conflicting as the nature of the absorption of vitamin B12 and the unique progression of the human digestive system as it has evolved over time Vitamin B12 was originally absorbed directly through plant sources and for some reason human digestive tracts developed the inability to do this and instead the absorption only became possible via the bacteria found in animals. This means that at some point, our bodies either ceased to carry this bacterium, we never carried it at all and had a method of synthesizing B12 on our own or animals themselves developed this process.

Perhaps this strange inconsistency (being that animals only carry the bacteria to synthesize B12) could be due to the introduction of meat into the diet seeing as humans were said to most likely originally be futilitarians and scavengers (consuming "left over" animal parts when they could). Often, as is evident in the obsolete nature of certain dormant organs such as the appendix, the body will intelligently re-design some of its mechanisms in order to adapt and refine itself for efficiency. If it was no longer or never necessary to carry a bacterium in order to synthesize B12 due to the consumption of animals, it is logical to assume that this could be a potential reason why we no longer or never carried it.

Although B12 is indeed absorbed through animal sources, it must be stressed that natural sources of B12 are found in soil fertilized by animals, animals that were here on Earth fertilizing soil long before us. Thus, giving them ample time to perfect the craft of B12 synthesization and leading to a common sense line of reasoning that animals were probably not the original source of the bacteria or B12, they instead became one over time which goes in accordance with the laws of evolution.

So how does this change the idea that eating animals is "natural" to the human diet and body? Currently there exists hardly enough evidence to support this claim, rather a large body of research that tells us a familiar story about human biology and our relationship to the environment.

The fact that B12 is only synthesized through bacteria found in animals when taken apart and examined piece by piece seems irrelevant in the face of the true matter: There is no predetermined "natural order" of the Earth, some animals eat other animals, some don't, some people eat other people, some just don't eat at all. If we are to solely use nature as a guide for how we live our lives in the modern human world then it will not only leave us confused but also allow for unending contradictions as there are exist multiple natural occurrences in the animal kingdom which our human morality and consciousness may find abhorrent: Such as rape, murder, cannibalism and incest. This is a fact that has been stressed in a multitude of ways, however, it is important to highlight when such examples (as the natural consumption processes of B12) are used normally just to suit one facet of an opinion or claim.

If there is a way to absorb B12 without eating animals and if B12 and said animals ancestors populated the Earth before we could even sprout a word, it seems illogical to automatically lead to the simple conclusion that animals are natural for us to consume . More likely so, soil is natural for us to eat as we have been ingesting its products on a more consistent basis before we could develop the tools to hunt. We have been ingesting the fruits of the soil long before we started eating the animals which we now eat at a vast percentage more. Today, we justify justifying this mass consumption by using a vitamin as a main point of argument.

While we may have scavenged and stumbled upon the occasional animal carcass in our early evolution, the point is that B12 is in the soil of which the carcass and all carcasses, even ours eventually become a part of. Meaning that when we process B12 we are getting it from the soil of which not only the animals we eat have discentigrated but also other animals which we would never eat such as dogs and even dinosaurs. Therefore, we can conclude that if eating animals is natural because we absorb B12 only from their manure, we must also adhere to the claim that it is natural to eat all of those other animals as well, even including HUMANS since our own manure will produce B12. Obviously, the nature of well-NATURE is that it is dynamic and often void of one dimensional explanations.

Historically and for most animals on Earth they have eaten what they could find and hunt. For humans, unfortunately, the Earth is our buffet which can lend itself to many hardened responses concerning vegan or vegetarianism. While we may have an overage in choices today, that was not the case for most of our past, we may not have had a choice but to eat animals in order to gain access to B12 in the past. Lacking a choice at one point may explain why we started eating meat in the first place but it does not entirely explain the intrinsic laws of nature as being set in stone.

Jasmonic Acid and its Related Compounds

Jasmonic acid (JA) is derived from the fatty acid linolenic acid. It is a member of the jasmonate class of plant hormones. It is biosynthesized from linolenic acid by the octadecanoid pathway.

The major function of JA and its various metabolites is regulating plant responses to abiotic and biotic stresses as well as plant growth and development.

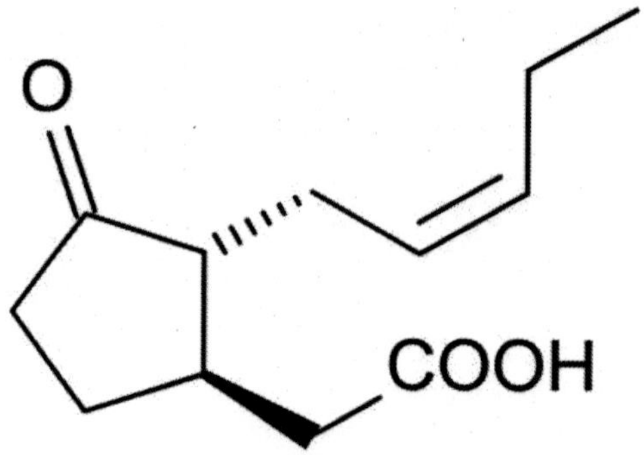

Regulated plant growth and development processes include growth inhibition, senescence, tendril coiling, flower development and leaf abscission. JA is also responsible for tuber formation in potatoes, yams, and onions.

It has an important role in response to wounding of plants and systemic acquired resistance. When plants are attacked by insects, they respond by releasing JA, which activates the expression of protease inhibitors, among many other anti-herbivore defence compounds. These protease inhibitors prevent proteolytic activity of the insects' digestive proteases or "salivary proteins", thereby stopping them from acquiring the needed nitrogen in the protein for their own growth.

Jasmonic acid is also converted to a variety of derivatives including esters such as methyl jasmonate; it may also be conjugated to amino acids.

This chemical may have a role in pest control, according to an October 2008 BBC News report. Researchers at the UK's Lancaster University have signed a licensing deal with an American company to market jasmonic acid as a seed treatment. JA seed treatment stimulates the natural anti-pest defences of the plants that germinate from the treated seeds, without harming plant growth as happens when jasmonates are sprayed onto plants that have already started growing.

Jasmonate

Jasmonate (JA) and its derivatives are lipid-based hormone signals that regulate a wide range of processes in plants, ranging from growth and photosynthesis to reproductive development. In particular, JAs are critical for plant defence against herbivory and plant responses to poor environmental conditions and other kinds of abiotic and biotic challenges. Some JAs can also be released as volatile organic compounds (VOCs) to permit communication between plants in anticipation of mutual dangers.

Figure: *Jasminum grandiflorum*

The isolation of methyl jasmonate from jasmine oil derived from *Jasminum grandiflorum* led to the discovery of the molecular structure of jasmonates and their name.

Structure of Jasmonates and Its Derivatives

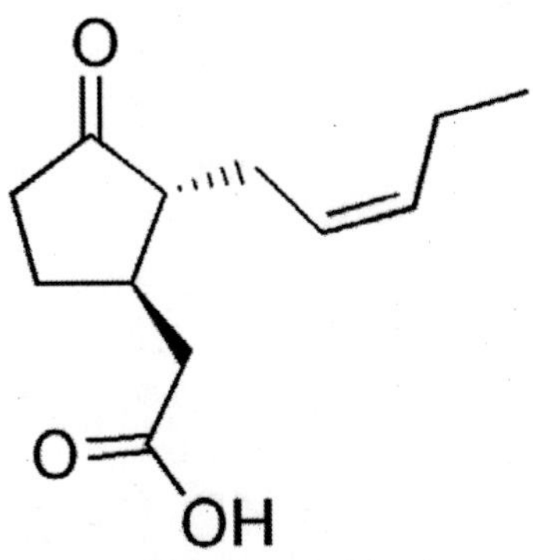

Jasmonic acid

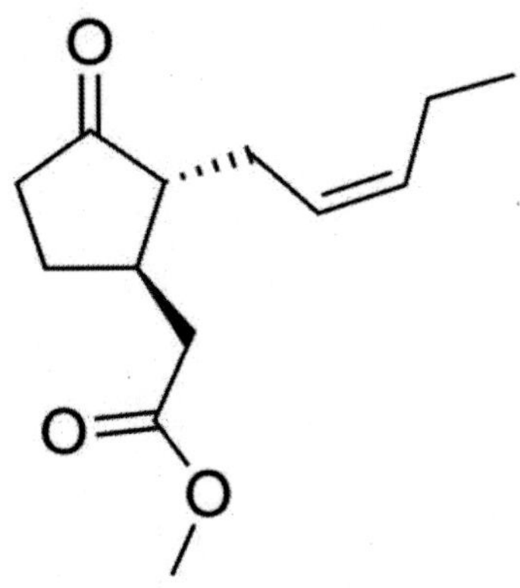

Methyl JA

Figure: *Structures of active jasmonate derivatives*

Jasmonates (JA) are an oxylipin, or oxygenated fatty acid, synthesized from linolenic acid residing in chloroplast membranes. Synthesis is initiated with the conversion of linolenic acid to 12-oxo-phytodienoic acid (OPDA), which then undergoes a reduction and three rounds of oxidation to form (+)-7-iso-JA, jasmonic acid proper. Only the conversion of linolenic acid to OPDA occurs in the chloroplast; all subsequent reactions occur in the peroxisome. JA itself can be further metabolized into active or inactive derivatives. Methyl JA (MeJA) is a volatile compound that is potentially responsible for interplant communication. JA conjugated with amino acid isoleucine (Ile) results in JA-Ile, which is currently the only known JA derivative needed for JA signaling.

Mechanism of Jasmonate Signaling

In general, the steps in jasmonate (JA) signaling mirror that of auxin signaling: the first step comprises E3 ubiquitin ligase complexes, which tag substrates with ubiquitin to mark them for degradation by proteasomes. The second step utilizes transcription factors to effect physiological changes.

One of the key molecules in this pathway is JAZ, which serves as the on-off switch for JA signaling. In the absence of JA, JAZ proteins bind to downstream transcription factors and limit their activity. However, in the presence of JA or its bioactive derivatives, JAZ proteins are degraded, freeing transcription factors for expression of genes needed in stress responses.

Because JAZ did not disappear in null *coi1* mutant plant backgrounds, protein COI1 was shown to mediate JAZ degradation. COI1 belongs to the family of highly conserved F-box proteins, and it recruits substrates for the E3 ubiquitin ligase SCF^{COI1}. The complexes that ultimately form are known as SCF complexes. These complexes bind JAZ and target it for proteasomal degradation. However, given the large spectrum of JA molecules, not all JA derivatives activate this pathway for signaling, and the range of those participating in this pathway is unknown. Thus far, only JA-Ile has been shown to be necessary for COI1-mediated degradation of JAZ11. JA-Ile and structurally related derivatives can bind to COI1-JAZ complexes and promote ubiquitination and thus degradation of the latter.

This mechanistic model raises the possibility that COI1 serves as an intracellular receptor for JA signals. Recent research has confirmed this hypothesis by demonstrating that the COI1-JAZ complex acts as a co-receptor for JA perception. Specifically, JA-Ile binds both to a ligand-binding pocket in COI1 and to a 20 amino-acid stretch of the conserved Jas motif in JAZ. This JAZ residue acts as a plug for the pocket in COI1, keeping JA-Ile bound in the pocket. Additionally, co-purification and subsequent removal of inositol pentakisphosphate ($InsP_5$) from COI1 suggest $InsP_5$ is a necessary component of the co-receptor and plays a role in potentiating the co-receptor complex.

Once freed from JAZ, transcription factors can activate genes needed for a specific JA response. The best-studied transcription factors acting in this pathway belong to the MYC family of transcription factors, which are characterized by a basic helix-loop-helix (bHLH) DNA binding motif. These factors (of which there are three, MYC2, 3, and 4) tend to act additively. For example, a plant that has only lost one myc becomes more susceptible to insect herbivory than a normal plant. A plant that has lost all three will be as susceptible to damage as coi1 mutants, which are completely unresponsive to JA and cannot mount a defence against herbivory. However, while all these MYC molecules share functions, they vary greatly in expression patterns and transcription functions. For instance, MYC2 has a greater effect on root growth compared to MYC3 or MYC4.

Additionally, MYC2 will loop back and regulate JAZ expression levels, leading to a negative feedback loop. These transcription factors all have different impacts on JAZ levels after JA signaling. JAZ levels in turn affect transcription factor and gene expression levels. In other words, on top of activating different response genes, the transcription factors can vary JAZ levels to achieve specificity in response to JA signals.

Roles of Jasmonate

Although jasmonate (JA) regulates many different processes in the plant, its role in wound response is best understood. Following mechanical wounding or herbivory, JA biosynthesis is rapidly activated, leading to expression of the appropriate response genes. For example, in the tomato, wounding produces defence molecules that inhibit leaf digestion in the insect's gut. Another indirect result of JA signaling is the volatile emission of JA-derived compounds. MeJA on leaves can travel airborne to nearby plants and elevate levels of transcripts related to wound response. In general, this emission can further upregulate JA synthesis and signaling and induce nearby plants to prime their defences in case of herbivory.

Following its role in defence, JAs have also been implicated in cell death and leaf senescence. JA can interact with many kinases and transcription factors associated with senescence. JA can also induce mitochondrial death by inducing the accumulation of reactive oxygen species (ROSs). These compounds disrupt mitochondria membranes and compromise the cell by causing apoptosis, or programmed cell death. JAs' roles in these processes are suggestive of methods by which the plant defends itself against biotic challenges and limits the spread of infections.

JA and its derivatives have also been implicated in plant development, symbiosis, and a host of other processes included in the list below.

- By studying mutants overexpressing JA, one of the earliest discoveries made was that JA inhibits root growth. The mechanism behind this event is still not understood, but mutants in the COI1-dependent signaling pathway tend to show reduced inhibition, demonstrating that the COI1 pathway is somehow necessary for inhibiting root growth.
- JA plays many roles in flower development. Mutants in JA synthesis or in JA signaling in Arabidopsis present with male sterility, typically due to delayed development. Interestingly, the same genes promoting male fertility in Arabidopsis promote female fertility in tomatoes. Overexpression of 12-OH-JA can also delay flowering.
- JA and MeJA inhibit the germination of nondormant seeds and stimulate the germination of dormant seeds.
- High levels of JA encourage the accumulation of storage proteins; genes encoding vegetative storage proteins are JA responsive. Specifically, tuberonic acid, a JA derivative, induces the formation of tubers.
- JAs also play a role in symbiosis between plants and microorganisms; however, its precise role is still unclear. JA currently appears to regulate signal exchange and nodulation regulation between legumes and rhizobium. On the other hand, elevated JA levels appear to regulate carbohydrate partitioning and stress tolerance in mycorrhizal plants.

Jasmonate and Pathogenesis

Pseudomonas syringae causes bacterial speck disease in tomatoes by hijacking the plant's jasmonate (JA) signaling pathway. This bacteria utilizes a type III secretion system to inject a cocktail of viral effector proteins into host cells.

One of the molecules included in this mixture is the phytotoxin coronatine (COR). JA-insensitive plants are highly resistant to *P. syringae* and unresponsive to COR; additionally, applying MeJA was sufficient to rescue virulence in COR mutant bacteria. Infected plants also expressed downstream JA and wound response genes but repressed levels of pathogenesis-related (PR) genes. All these data suggest COR acts through the JA pathway to invade host plants. Activation of a wound response is hypothesized to come at the expense of pathogen defence. By activating the JA wound response pathway, *P. syringae* could divert resources from its host's immune system and infect more effectively.

Cross Talk with Other Defence Pathways

While the jasmonate (JA) pathway is critical for wound response, it is not the only signaling pathway mediating defence in plants. To build an optimal yet efficient defence, the different defence pathways must be capable of cross talk to fine-tune and specify responses to abiotic and biotic challenges.

One of the best studied examples of JA cross talk occurs with salicylic acid (SA). SA, a hormone, mediates defence against pathogens by inducing both the expression of pathogenesis-related genes and systemic acquired resistance (SAR), in which the whole plant gains resistance to a pathogen after localized exposure to it.

Wound and pathogen response appear to be interact negatively. For example, silencing phenylalanine ammonia lyase (PAL), an enzyme synthesizing precursors to SA, reduces SAR but enhances herbivory resistance against insects. Similarly, overexpression of PAL enhances SAR but reduces wound response after insect herbivory. Generally, it has been found that pathogens living in live plant cells are more sensitive to SA-induced defences, while herbivorous insects and pathogens that derive benefit from cell death are more susceptible to JA defences. Thus, this trade-off in pathways optimizes defence and saves plant resources.

Cross talk also occurs between JA and other plant hormone pathways, such as those of abscisic acid (ABA) and ethylene (ET). These interactions similarly optimize defence against pathogens and herbivores of different lifestyles. For example, MYC2 activity can be stimulated by both JA and ABA pathways, allowing it to integrate signals from both pathways. Other transcription factors such as ERF1 arise as a result of JA and ET signaling. All these molecules can act in combination to activate specific wound response genes.

Finally, cross talk is not restricted for defence: JA and ET interactions are critical in development as well, and a balance between the two compounds is necessary for proper apical hook development in *Arabidopsis* seedlings. Still, further research is needed to elucidate the molecules regulating such cross talk.

Oxylipin

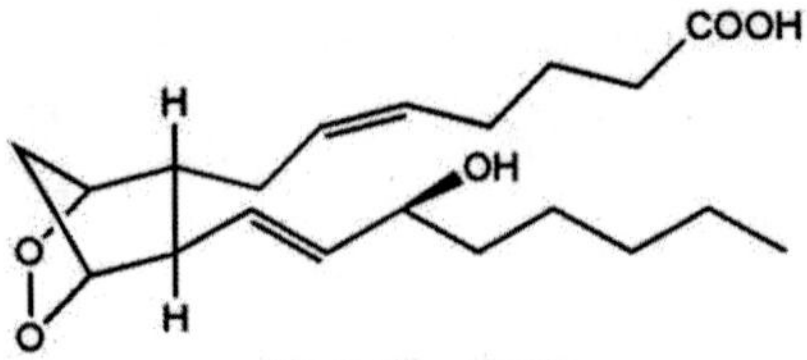

Prostaglandin H_2

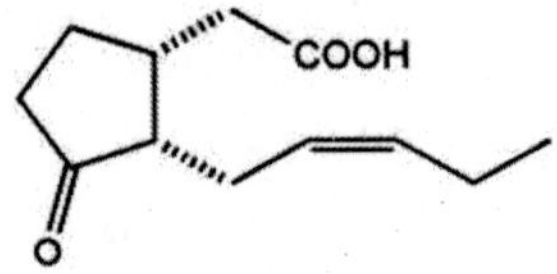

(+)-7-iso-Jasmonic acid

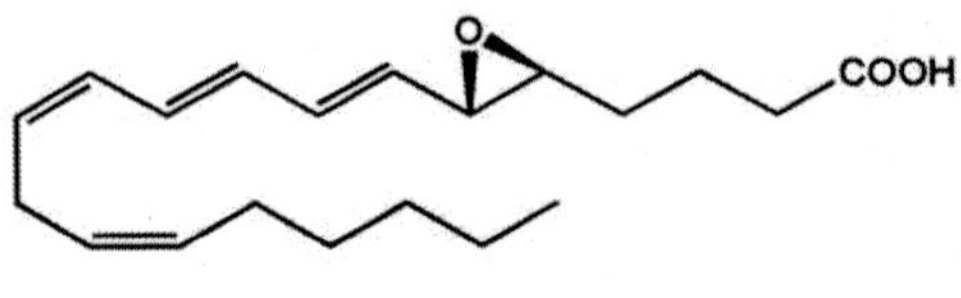

Leukotriene A_4

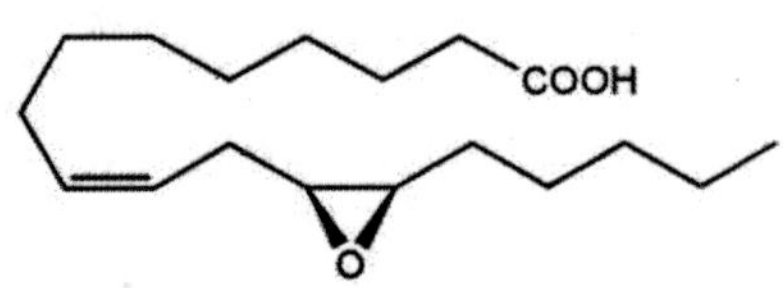

(+)-Vernolic acid

Figure: *The structural formulae of selected oxylipins*

Oxylipins constitute a family of oxygenated natural products which are formed from fatty acids by pathways involving at least one step of dioxygen-dependent oxidation. Many of oxylipins have physiological significance.

Oxylipins are widespread in aerobic organisms including plants, animals and fungi. Typically, oxylipins are not stored in tissues but are formed on demand by liberation of precursor fatty acids from esterified forms.

Biosynthesis of oxylipins is initiated by dioxygenases or monooxygenases; however also non-enzymatic autoxidative processes contribute to oxylipin formation (phytoprostanes, isoprostanes). Dioxygenases include lipoxygenases (plants, animals, fungi), heme-dependent fatty acid oxygenases (plants, fungi), and cyclooxygenases (animals). Fatty acid hydroperoxides or endoperoxides are formed by action of these enzymes. Monooxygenases involved in oxylipin biosynthesis are members of the cytochrome P450 superfamily and can oxidize double bonds with epoxide formation or saturated carbons forming alcohols. Nature has evolved numerous enzymes which metabolize oxylipins into secondary products, many of which possess strong biological activity. Of special importance are the cytochrome P450 enzymes in animals, including CYP5A1 (thromboxane synthase), CYP8A1 (prostacyclin synthase), and the CYP74 family of hydroperoxide-metabolizing enzymes in plants, lower animals and bacteria. In the plant and animal kingdoms the C18 and C20 polyenoic fatty acids, respectively, are the major precursors of oxylipins.

Oxylipins in animals, referred to as eicosanoids (Greek icosa; twenty) because of their formation from twenty-carbon essential fatty acids, have potent and often opposing effects on e.g. smooth muscle (vasculature, myometrium) and blood platelets. Certain eicosanoids (leukotrienes B4 and C4) are proinflammatory whereas others (resolvins, protectins) are antiinflammatory and are involved in the resolution process which follows tissue injury. Plant oxylipins are mainly involved in control of ontogenesis, reproductive processes and in the resistance to various microbial pathogens and other pests.

8

Nitrogen and Phosphorus

Nitrogen is an essential building block of amino and nucleic acids, essential to life on Earth.

Elemental nitrogen in the atmosphere cannot be used directly by either plants or animals, and must be converted to a reduced (or 'fixed') state to be useful for higher plants and animals. Precipitation often contains substantial quantities of ammonium and nitrate, thought to result from nitrogen fixation by lightning and other atmospheric electric phenomena.

This was first proposed by Liebig in 1827 and later confirmed. However, because ammonium is preferentially retained by the forest canopy relative to atmospheric nitrate, most fixed nitrogen reaches the soil surface under trees as nitrate. Soil nitrate is preferentially assimilated by tree roots relative to soil ammonium.

Specific bacteria (e.g., *Rhizobium trifolium*) possess nitrogenase enzymes that can fix atmospheric nitrogen into a form (ammonium ion) that is chemically useful to higher organisms. This process requires a large amount of energy and anoxic conditions. Such bacteria may live freely in soil (e.g., *Azotobacter*) but normally exist in a symbiotic relationship in the root nodules of leguminous plants (e.g. clover, *Trifolium*, or soybean plant, *Glycine max*) and fertilizer trees. Nitrogen-fixing bacteria are also symbiotic with a number of unrelated plant species such as alders (*Alnus*) spp., lichens, *Casuarina*, *Myrica*, liverworts, and *Gunnera*.

As part of the symbiotic relationship, the plant converts the 'fixed' ammonium ion to nitrogen oxides and amino acids to form proteins and other molecules, (e.g., alkaloids). In return for the 'fixed' nitrogen, the plant secretes sugars to the symbiotic bacteria. Legumes maintain an anaerobic (oxygen free) environment for their nitrogen-fixing bacteria.

Plants are able to assimilate nitrogen directly in the form of nitrates that may be present in soil from natural mineral deposits, artificial fertilizers, animal waste, or organic decay (as the product of bacteria, but not bacteria specifically associated with the plant). Nitrates absorbed in this fashion are converted to nitrites by the enzyme *nitrate* reductase, and then converted to ammonia by another enzyme called *nitrite* reductase.

Nitrogen compounds are basic building blocks in animal biology as well. Animals use nitrogen-containing amino acids from plant sources as starting materials for all nitrogen-compound animal biochemistry, including the manufacture of proteins and nucleic acids. Plant-feeding insects are dependent on nitrogen in their diet, such that varying the amount of nitrogen fertilizer applied to a plant can affect the reproduction rate of insects feeding on fertilized plants.

Soluble nitrate is an important limiting factor in the growth of certain bacteria in ocean waters. In many places in the world, artificial fertilizers applied to crop-lands to increase yields result in run-off delivery of soluble nitrogen to oceans at river mouths. This process can result in eutrophication of the water, as nitrogen-driven bacterial growth depletes water oxygen to the point that all higher organisms die. Well-known "dead zone" areas in the U.S. Gulf Coast and the Black Sea are due to this important polluting process.

Many saltwater fish manufacture large amounts of trimethylamine oxide to protect them from the high osmotic effects of their environment; conversion of this compound to dimethylamine is responsible for the early odor in unfresh saltwater fish. In animals, free radical nitric oxide (NO) (derived from an amino acid), serves as an important regulatory molecule for circulation.

Nitric oxide's rapid reaction with water in animals results in production of its metabolite nitrite. Animal metabolism of nitrogen in proteins, in general, results in excretion of urea, while animal metabolism of nucleic acids results in excretion of urea and uric acid. The characteristic odor of animal flesh decay is caused by the creation of long-chain, nitrogen-containing amines, such as putrescine and cadaverine, which are breakdown products of the amino acids ornithine and lysine, respectively, in decaying proteins.

Decay of organisms and their waste products may produce small amounts of nitrate, but most decay eventually returns nitrogen content to the atmosphere, as molecular nitrogen. The circulation of nitrogen from atmosphere, to organic compounds, then back to the atmosphere, is referred to as the nitrogen cycle.

Nitrogen Cycle

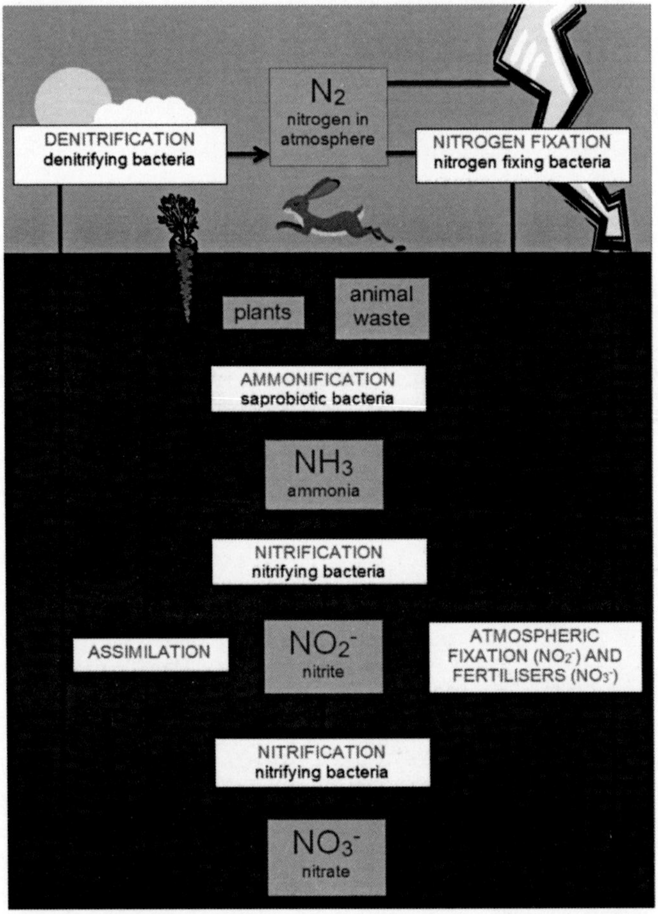

Figure: *A simple diagram of the nitrogen cycle. The blue boxes represent stores of nitrogen, the green writing is for processes that occur to move the nitrogen from one place to another and the red writing are all the bacteria involved.*

The nitrogen cycle is the process by which nitrogen is converted between its various chemical forms. This transformation can be carried out through both biological and physical processes. Important processes in the nitrogen cycle include fixation, ammonification, nitrification, and denitrification. The majority of Earth's atmosphere (78%) is nitrogen, making it the largest pool of nitrogen. However, atmospheric nitrogen has limited availability for biological use, leading to a scarcity of usable nitrogen in many types of ecosystems. The nitrogen cycle is of particular interest to ecologists because nitrogen availability can affect the rate of key ecosystem processes, including primary production and decomposition. Human activities such as fossil fuel combustion, use of artificial nitrogen fertilizers, and release of nitrogen in wastewater have dramatically altered the global nitrogen cycle.

Ecological Function

Nitrogen is necessary for all known forms of life on Earth. It is a component in all amino acids, as incorporated into proteins, and is present in the bases that make up nucleic acids such as RNA and DNA. In plants, much of the nitrogen is used in chlorophyll molecules, which are essential for photosynthesis and further growth. Nitrogen gas (N_2) is the largest constituent of the Earth's atmosphere, but this form is relatively nonreactive and unusable by plants. Chemical processing or natural fixation (through processes such as bacterial conversion—see rhizobium) are necessary to convert gaseous nitrogen into compounds such as nitrate or ammonia which can be used by plants. The abundance or scarcity of this "fixed" nitrogen (also known as reactive nitrogen) frequently limits plant growth in both managed and wild environments. The nitrogen cycle, like the carbon cycle, is an important part of every ecosystem.

The Processes of the Nitrogen Cycle

Nitrogen is present in the environment in a wide variety of chemical forms including organic nitrogen, ammonium (NH_4^+), nitrite (NO_2^-), nitrate (NO_3^-), nitrous oxide (N_2O), nitric oxide (NO) or inorganic nitrogen gas (N_2). Organic nitrogen may be in the form of a living organism, humus or in the intermediate products of organic matter decomposition. The processes of the nitrogen cycle transform nitrogen from one form to another. Many of those processes are carried out by microbes, either in their effort to harvest energy or to accumulate nitrogen in a form needed for their growth. The diagram above shows how these processes fit together to form the nitrogen cycle.

Nitrogen Fixation

Atmospheric nitrogen must be processed, or "fixed", to be used by plants. Some fixation occurs in lightning strikes, but most fixation is done by free-living or symbiotic bacteria known as diazotrophs. These bacteria have the nitrogenase enzyme that combines gaseous nitrogen with hydrogen to produce ammonia, which is converted by the bacteria into other organic compounds. Most biological nitrogen fixation occurs by the activity of Mo-nitrogenase, found in a wide variety of bacteria and some Archaea. Mo-nitrogenase is a complex two component enzyme that has multiple metal-containing prosthetic groups. An example of the free-living bacteria is *Azotobacter*. Symbiotic nitrogen-fixing bacteria such as *Rhizobium* usually live in the root nodules of legumes (such as peas, alfalfa, and locust trees). Here they form a mutualistic relationship with the plant, producing ammonia in exchange for carbohydrates. Because of this relationship, legumes will often increase the nitrogen content of nitrogen-poor soils. A few non-legumes can also form such symbioses. Today, about 30% of the total fixed nitrogen is produced industrially using the Haber-Bosch process, which uses high temperatures and pressures to convert nitrogen gas and a hydrogen source (natural gas or petroleum) into ammonia.

Biological Nitrogen Fixation

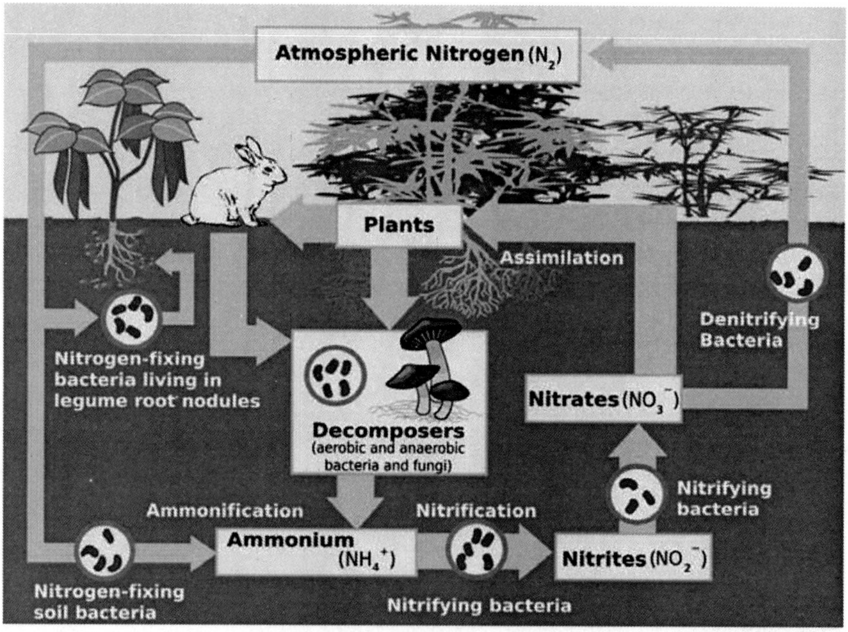

Figure: *Schematic representation of the nitrogen cycle. Abiotic nitrogen fixation has been omitted.*

Biological nitrogen fixation was discovered by the German agronomist Hermann Hellriegel and Dutch microbiologist Martinus Beijerinck. Biological nitrogen fixation (BNF) occurs when atmospheric nitrogen is converted to ammonia by an enzyme called nitrogenase. The reaction for BNF is:

$$N_2 + 8\ H^+ + 8\ e^- \rightarrow 2\ NH_3 + H_2$$

The process is coupled to the hydrolysis of 16 equivalents of ATP and is accompanied by the co-formation of one molecule of H_2. In free-living diazotrophs, the nitrogenase-generated ammonium is assimilated into glutamate through the glutamine synthetase/glutamate synthase pathway. The microbial genes required for nitrogen fixation are widely distributed in diverse environments.

Enzymes responsible for nitrogenase action are very susceptible to destruction by oxygen. Many bacteria cease production of the enzyme in the presence of oxygen. Many nitrogen-fixing organisms exist only in anaerobic conditions, respiring to draw down oxygen levels, or binding the oxygen with a protein such as leghemoglobin.

Microorganisms that Fix Nitrogen

Diazotrophs are cyanobacteria, e.g. the highly significant trichodesmium , green sulfur bacteria, azotobacteraceae, rhizobia and Frankia.

Cyanobacteria inhabit nearly all illuminated environments on Earth and play key roles in the carbon and nitrogen cycle of the biosphere. In general, cyanobacteria are able to utilize a variety of inorganic and organic sources of combined nitrogen, like nitrate, nitrite, ammonium, urea, or some amino acids. Several cyanobacterial strains are also capable of diazotrophic growth, an ability that may have been present in their last common ancestor in the Archaean. Nitrogen fixation by cyanobacteria in coral reefs can fix twice the amount of nitrogen than on land—around 1.8 kg of nitrogen is fixed per hectare per day. The colonial marine cyanobacterium *Trichodesmium* is thought to fix nitrogen on such a scale that it accounts for almost half of the nitrogen-fixation in marine systems on a global scale.

Root Nodule Symbioses

Legume Family: Plants that contribute to nitrogen fixation include the legume family – Fabaceae – with taxa such as kudzu, clovers, soybeans, alfalfa, lupines, peanuts, and rooibos. They contain symbiotic bacteria called *Rhizobia* within nodules in their root systems, producing nitrogen compounds that help the plant to grow and compete with other plants. When the plant dies, the fixed nitrogen is released, making it available to other plants and this helps to fertilize the soil. The great majority of legumes have this association, but a few genera (e.g., *Styphnolobium*) do not. In many traditional and organic farming practices, fields are rotated through various types of crops, which usually includes one consisting mainly or entirely of clover or buckwheat (non-legume family *Polygonaceae*), which are often referred to as "green manure". Inga alley farming relies on the leguminous genus *Inga*, a small tropical, tough-leaved, nitrogen-fixing tree.

Non-leguminous

Although by far the majority of plants able to form nitrogen-fixing root nodules are in the legume family Fabaceae, there are a few exceptions:

- *Parasponia*, a tropical genus in the Cannabaceae also able to interact with rhizobia and form nitrogen-fixing nodules
- Actinorhizal plants such as alder and bayberry can also form nitrogen-fixing nodules, thanks to a symbiotic association with *Frankia* bacteria. These plants belong to 25 genera distributed among 8 plant families.

The ability to fix nitrogen is far from universally present in these families. For instance, of 122 genera in the Rosaceae, only 4 genera are capable of fixing nitrogen. All these families belong to the orders Cucurbitales, Fagales, and Rosales, which together with the Fabales form a clade of eurosids. In this clade, Fabales were the first lineage to branch off; thus, the ability to fix nitrogen may be plesiomorphic and subsequently lost in most descendants of the original nitrogen-fixing plant; however, it may be that the basic genetic and physiological requirements were present in an incipient state in the last common ancestors of all these plants, but only evolved to full function in some of them.

Assimilation

Biological assimilation, or *bio assimilation*, is the combination of two processes to supply animal cells with nutrients. The first is the process of absorbing vitamins, minerals, and other chemicals from food within the gastrointestinal tract. In humans this is done with a chemical breakdown (enzymes and acids) and physical breakdown (oral mastication and stomach churning.) The second process of bio assimilation is the chemical alteration of substances in the bloodstream by the liver or cellular secretions. Although a few similar compounds can be absorbed in digestion bio assimilation, the bioavailability of many compounds is dictated by this second process since both the liver and cellular secretions can be very specific in their metabolic action. This second process is where the absorbed food reaches the cells via the liver.

Plants take nitrogen from the soil by absorption through their roots in the form of either nitrate ions or ammonium ions. Most nitrogen obtained by terrestrial animals can be traced back to the eating of plants at some stage of the food chain.

Plants can absorb nitrate or ammonium ions from the soil via their root hairs. If nitrate is absorbed, it is first reduced to nitrite ions and then ammonium ions for incorporation into amino acids, nucleic acids, and chlorophyll. In plants that have a symbiotic relationship with rhizobia, some nitrogen is assimilated in the form of ammonium ions directly from the nodules. It is now known that there is a more complex cycling of amino acids between *Rhizobia* bacteroids and plants. The plant provides amino acids to the bacteroids so ammonia assimilation is not required and the bacteroids pass amino acids (with the newly fixed nitrogen) back to the plant, thus forming an interdependent relationship. While many animals, fungi, and other heterotrophic organisms obtain nitrogen by ingestion of amino acids, nucleotides and other small organic molecules, other heterotrophs (including many bacteria) are able to utilize inorganic compounds, such as ammonium as sole N sources. Utilization of various N sources is carefully regulated in all organisms.

Ammonification

When a plant or animal dies or an animal expels waste, the initial form of nitrogen is organic. Bacteria or fungi convert the organic nitrogen within the remains back into ammonium (NH_4^+), a process called ammonification or mineralization. Enzymes involved are:

- GS: Gln Synthetase (Cytosolic & PLastid)
- GOGAT: Glu 2-oxoglutarate aminotransferase (Ferredoxin & NADH dependent)
- GDH: Glu Dehydrogenase:
  - o Minor Role in ammonium assimilation.
  - o Important in amino acid catabolism.

Nitrification

Nitrification is the biological oxidation of ammonia with oxygen, then into ammonium, then into nitrite followed by the oxidation of these nitrites into nitrates. Degradation of ammonia to nitrite is usually the rate limiting step of nitrification. Nitrification is an important step in the nitrogen cycle in soil. Nitrification is an aerobic process performed by small groups of autotrophic bacteria and archaea.

Microbiology and Ecology

The oxidation of ammonia into nitrite is performed by two groups of organisms, ammonia-oxidizing bacteria (AOB) and ammonia-oxidizing archaea (AOA). AOB can be found among the β-proteobacteria and gammaproteobacteria. Currently, two AOA, *Nitrosopumilus maritimus* and 'Nitrososphaera viennensis", have been isolated and described. In soils the most studied AOB belong to the genera *Nitrosomonas* and *Nitrosococcus*. Although in soils ammonia oxidation occurs by both AOB and AOA, AOA dominate in both soils and marine environments, suggesting that *Thaumarchaeota* may be greater contributors to ammonia oxidation in these environments.

The second step (oxidation of nitrite into nitrate) is done (mainly) by bacteria of the genus *Nitrobacter* and "Nitrospira". Both steps are producing energy to be coupled to ATP synthesis. Nitrifying organisms are chemoautotrophs, and use carbon dioxide as their carbon source for growth. Some AOB possess the enzyme, urease, which catalyzes the conversion of the urea molecule to two ammonia molecules and one carbon dioxide molecule. *Nitrosomonas europaea*, as well as populations of soil-dwelling AOB, have been shown to assimilate the carbon dioxide released by the reaction to make biomass via the Calvin Cycle, and harvest energy by oxidizing ammonia (the other product of urease) to nitrite. This feature may explain enhanced growth of AOB in the presence of urea in acidic environments.

In most environments, organisms are present that will complete both steps of the process, yielding nitrate as the final product. However, it is possible to design systems in which nitrite is formed (the *Sharon process*).

Nitrification is important in agricultural systems, where fertilizer is often applied as ammonia. Conversion of this ammonia to nitrate increases nitrogen leaching because nitrate is more water-soluble than ammonia.

Nitrification also plays an important role in the removal of nitrogen from municipal wastewater. The conventional removal is nitrification, followed by denitrification. The cost of this process resides mainly in aeration (bringing oxygen in the reactor) and the addition of an external carbon source (e.g., methanol) for the denitrification.

Nitrification can also occur in drinking water. In distribution systems where chloramines are used as the secondary disinfectant, the presence of free ammonia can act as a substrate for ammonia-oxidizing microorganisms. The associated reactions can lead to the depletion of the disinfectant residual in the system. The addition of chlorite ion to chloramine-treated water has been shown to control nitrification.

Together with ammonification, nitrification forms a mineralization process that refers to the complete decomposition of organic material, with the release of available nitrogen compounds. This replenishes the nitrogen cycle.

Chemistry

Nitrification is a process of nitrogen compound oxidation (effectively, loss of electrons from the nitrogen atom to the oxygen atoms):

1. $2\ NH_4^+ + 3\ O_2 \rightarrow 2\ NO_2^- + 2\ H_2O + 4\ H^+$ (Nitrosomonas)
2. $2\ NO_2^- + 1\ O_2 \rightarrow 2\ NO_3^-$ (Nitrobacter, Nitrospina)
3. $NH_3 + O_2 \rightarrow NO_2^- + 3H^+ + 2e^-$
4. $NO_2^- + H_2O \rightarrow NO_3^- + 2H^+ + 2e^-$

Nitrification in the marine environment

In the marine environment, nitrogen is often the limiting nutrient, so the nitrogen cycle in the ocean is of particular interest. The nitrification step of the cycle is of particular interest in the ocean because it creates nitrate, the primary form of nitrogen responsible for "new" production. Furthermore, as the ocean becomes enriched in anthropogenic CO_2, the resulting decrease in pH could lead to decreasing rates of nitrification. Nitrification could potentially become a "bottleneck" in the nitrogen cycle.

Nitrification, as stated above, is formally a two-step process; in the first step ammonia is oxidized to nitrite, and in the second step nitrite is oxidized to nitrate. Different microbes are responsible for each step in the marine environment. Several groups of ammonia-oxidizing bacteria (AOB) are known in the marine environment, including *Nitrosomonas*, *Nitrospira*, and *Nitrosococcus*. All contain the functional gene ammonia monooxygenase (AMO) which, as its name implies, is responsible for the oxidation of ammonia. More recent metagenomic studies have revealed that some Thaumarchaeota (formerly Crenarchaeota) possess AMO. Thaumarchaeotes are abundant in the ocean and some species have a 200 times greater affinity for ammonia than AOB, leading researchers to challenge the previous belief that AOB are primarily responsible for nitrification in the ocean. Furthermore, though nitrification is classically thought to be vertically separated from primary production because the oxidation of nitrogen by bacteria is inhibited by light, nitrification by AOA does not appear to be light inhibited, meaning that nitrification is occurring throughout the water column, challenging the classical definitions of "new" and "recycled" production.

In the second step, nitrite is oxidized to nitrate. In the ocean, this step is not as well understood as the first, but the bacteria *Nitrospina* and *Nitrobacter* are known to carry out this step in the ocean.

Soil Conditions Controlling Nitrification Rates

- Substrate availability (presence of NH_4^+)
- Aeration (availability of O2)
- Well-drained soils with 60% soil moisture
- pH (near neutral)
- Temperature (best 20-30°C) => Nitrification is seasonal, affected by land use practices

Denitrification

Denitrification is the reduction of nitrates back into the largely inert nitrogen gas (N_2), completing the nitrogen cycle. This process is performed by bacterial species such as *Pseudomonas* and *Clostridium* in anaerobic conditions. They use the nitrate as an electron acceptor in the place of oxygen during respiration. These facultatively anaerobic bacteria can also live in aerobic conditions. Denitrification happens in anaerobic conditions e.g. waterlogged soils. The denitrifying bacteria use nitrates in the soil to carry out respiration and consequently produce nitrogen gas, which is inert and unavailable to plants.

Anaerobic Ammonium Oxidation

In this biological process, nitrite and ammonium are converted directly into molecular nitrogen (N_2) gas. This process makes up a major proportion of nitrogen conversion in the oceans.

Other Processes

Though nitrogen fixation is the primary source of plant-available nitrogen in most ecosystems, in areas with nitrogen-rich bedrock, the breakdown of this rock also serves as a nitrogen source.

Marine Nitrogen Cycle

The nitrogen cycle is an important process in the ocean as well. While the overall cycle is similar, there are different players and modes of transfer for nitrogen in the ocean. Nitrogen enters the water through precipitation, runoff, or as N_2 from the atmosphere. Nitrogen cannot be utilized by phytoplankton as N_2 so it must undergo nitrogen fixation which is performed predominately by cyanobacteria. Without supplies of fixed nitrogen entering the marine cycle the fixed nitrogen would be used up in about 2000 years. Phytoplankton need nitrogen in biologically available forms for the initial synthesis of organic matter. Ammonia and urea are released into the water by excretion from plankton. Nitrogen sources are removed from the euphotic zone by the downward movement of the organic matter. This can occur from sinking of phytoplankton, vertical mixing, or sinking of waste of vertical migrators. The sinking results in ammonia being introduced at lower depths below the euphotic zone. Bacteria are able to convert ammonia to nitrite and nitrate but they are inhibited by light so this must occur below the euphotic zone. Ammonification or Mineralization is performed by bacteria to convert the ammonia to ammonium. Nitrification can then occur to convert the ammonium to nitrite and nitrate. Nitrate can be returned to the euphotic zone by vertical mixing and upwelling where it can be taken up by phytoplankton to continue the cycle. N_2 can be returned to the atmosphere through denitrification.

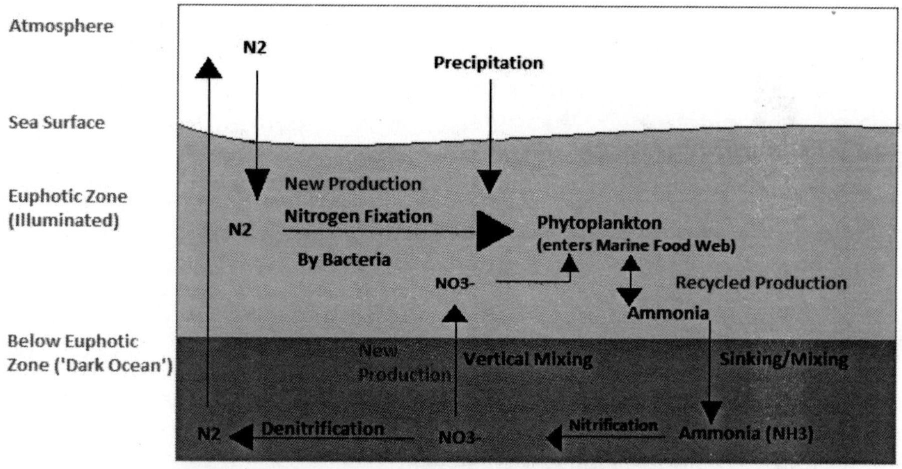

Figure: *A schematic representing the Marine Nitrogen Cycle*

Ammonium is thought to be the preferred source of fixed nitrogen for phytoplankton because its assimilation does not involve a redox reaction and therefore requires little energy. Nitrate requires a redox reaction for assimilation, but is more abundant so most phytoplankton have adapted to have the enzymes necessary to undertake this reduction (nitrate reductase). There are a few notable and well-known exceptions that include Prochlorococcus and some Synechococcus. These species can only take up nitrogen as ammonium.

The nutrients in the ocean are not uniformly distributed. Areas of upwelling provide supplies of nitrogen from below the euphotic zone. Coastal zones provide nitrogen from runoff and upwelling occurs readily along the coast. However, the rate at which nitrogen can be taken up by phytoplankton is decreased in oligotrophic waters year-round and temperate water in the summer resulting in lower primary production. The distribution of the different forms of nitrogen varies throughout the oceans as well.

Nitrate is depleted in near-surface water except in upwelling regions. Coastal upwelling regions usually have high nitrate and chlorophyll levels as a result of the increased production. However, there are regions of high surface nitrate but low chlorophyll that are referred to as HNLC (high nitrogen, low chlorophyll) regions. The best explanation for HNLC regions relates to iron scarcity in the ocean, which may play an important part in ocean dynamics and nutrient cycles. The input of iron varies by region and is delivered to the ocean by dust (from dust storms) and leached out of rocks. Iron is under consideration as the true limiting element to ecosystem productivity in the ocean.

Ammonium and nitrite show a maximum concentration at 50–80 m (lower end of the euphotic zone) with decreasing concentration below that depth. This distribution can be accounted for by the fact that nitrite and ammonium are intermediate species. They are both rapidly produced and consumed through the water column. The amount of ammonium in the ocean is about 3 orders of magnitude less than nitrate. Between ammonium, nitrite, and nitrate, nitritie has the fastest turnover rate. It can be produced during nitrate assimilation, nitrification, and denitrification; however, it is immediately consumed again.

New vs. Regenerated Nitrogen

Nitrogen entering the euphotic zone is referred to as new nitrogen because it is newly arrived from outside the productive layer. The new nitrogen can come from below the euphotic zone or from outside sources. Outside sources are upwelling from deep water and nitrogen fixation. If the organic matter is eaten, respired, delivered to the water as ammonia, and re-incorporated into organic matter by phytoplankton it is considered recycled/regenerated production.

New production is an important component of the marine environment. One reason is that only continual input of new nitrogen can determine the total capacity of the ocean to produce a sustainable fish harvest. Harvesting fish from regenerated nitrogen areas will lead to a decrease in nitrogen and therefore a decrease in primary production. This will have a negative effect on the system. However, if fish are harvested from areas of new nitrogen the nitrogen will be replenished.

Human Influences on the Nitrogen Cycle

As a result of extensive cultivation of legumes (particularly soy, alfalfa, and clover), growing use of the Haber–Bosch process in the creation of chemical fertilizers, and pollution emitted by vehicles and industrial plants, human beings have more than doubled the annual transfer of nitrogen into biologically available forms. In addition, humans have significantly contributed to the transfer of nitrogen trace gases from Earth to the atmosphere and from the land to aquatic systems. Human alterations to the global nitrogen cycle are most intense in developed countries and in Asia, where vehicle emissions and industrial agriculture are highest. Nitrous oxide (N_2O) has risen in the atmosphere as a result of agricultural fertilization, biomass burning, cattle and feedlots, and industrial sources. N_2O has deleterious effects in the stratosphere, where it breaks down and acts as a catalyst in the destruction of atmospheric ozone. Nitrous oxide is also a greenhouse gas and is currently the third largest contributor to global warming, after carbon dioxide and methane. While not as abundant in the atmosphere as carbon dioxide, it is, for an equivalent mass, nearly 300 times more potent in its ability to warm the planet.

Ammonia (NH_3) in the atmosphere has tripled as the result of human activities. It is a reactant in the atmosphere, where it acts as an aerosol, decreasing air quality and clinging to water droplets, eventually resulting in nitric acid (HNO_3) that produces acid rain. Atmospheric ammonia and nitric acid also damage respiratory systems.

The very-high temperature of lightning naturally produces small amounts of NO_x, NH_3, and HNO_3, but high-temperature combustion has contributed to a 6 or 7 fold increase in the flux of NO_x to the atmosphere. Its production is a function of combustion temperature - the higher the temperature, the more NO_x is produced. Fossil fuel combustion is a primary contributor, but so are biofuels and even the burning of hydrogen. The higher combustion temperature of hydrogen produces more NO_x than natural gas combustion.

Ammonia and nitrous oxides actively alter atmospheric chemistry. They are precursors of tropospheric (lower atmosphere) ozone production, which contributes to smog and acid rain, damages plants and increases nitrogen inputs to ecosystems. Ecosystem processes can increase with nitrogen fertilization, but anthropogenic input can also result in nitrogen saturation, which weakens productivity and can damage the health of plants, animals, fish, and humans.

Decreases in biodiversity can also result if higher nitrogen availability increases nitrogen-demanding grasses, causing a degradation of nitrogen-poor, species diverse heathlands.

Wastewater Treatment

Onsite sewage facilities such as septic tanks and holding tanks release large amounts of nitrogen into the environment by discharging through a drainfield into the ground. Microbial activity consumes the nitrogen and other contaminants in the wastewater. However, in certain areas, microbial activity is unable to process all the contaminants and some or all of the wastewater, with the contaminants, enters the aquifers. These contaminants accumulate and eventually end up in drinking water. One of the contaminants most concerned about is nitrogen in the form of nitrate. A nitrate concentration of 10 ppm (parts per million) or 10 milligrams per litre is the current EPA limit for drinking water and typical household wastewater can produce a range of 20–85 ppm.

One health risk associated with drinking water (with >10 ppm nitrate) is the development of methemoglobinemia and has been found to cause blue baby syndrome. Several American states have now started programmes to introduce advanced wastewater treatment systems to the typical onsite sewage facilities. The result of these systems is an overall reduction of nitrogen, as well as other contaminants in the wastewater.

Environmental Impacts

Additional risks posed by increased availability of inorganic nitrogen in aquatic ecosystems include water acidification; eutrophication of fresh and saltwater systems; and toxicity issues for animals, including humans. Eutrophication often leads to lower dissolved oxygen levels in the water column, including hypoxic and anoxic conditions, which can cause death of aquatic fauna. Relatively sessile benthos, or bottom-dwelling creatures, are particularly vulnerable because of their lack of mobility, though large fish kills are not uncommon. Oceanic dead zones near the mouth of the Mississippi in the Gulf of Mexico are a well-known example of algal bloom-induced hypoxia. The New York Adirondack Lakes, Catskills, Hudson Highlands, Rensselaer Plateau and parts of Long Island display the impact of nitric acid rain deposition, resulting in the killing of fish and many other aquatic species.

Ammonia (NH_3) is highly toxic to fish and the level of ammonia discharged from wastewater treatment facilities must be closely monitored. To prevent fish deaths, nitrification via aeration prior to discharge is often desirable. Land application can be an attractive alternative to the aeration.

Nitrate

Nitrate is a polyatomic ion with the molecular formula NO_3^- and a molecular mass of 62.0049 g/mol. Nitrates also describe the organic functional group $RONO_2$. These nitrate esters are a specialized class of explosives.

Structure

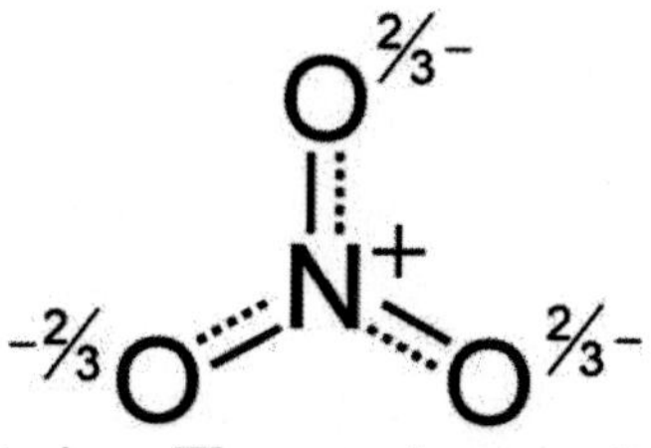

Figure: *The nitrate ion. The net charge of the whole ion is 1⁻.*

The anion is the conjugate base of nitric acid, consisting of one central nitrogen atom surrounded by three identically bonded oxygen atoms in a trigonal planar arrangement. The nitrate ion carries a formal charge of -1. This results from a combination formal charge in which each of the three oxygens carries a $-^2/_3$ charge, whereas the nitrogen carries a 1+ charge, all these adding up to formal charge of the polyatomic nitrate ion. This arrangement is commonly used as an example of resonance. Like the isoelectronic carbonate ion, the nitrate ion can be represented by resonance structures:

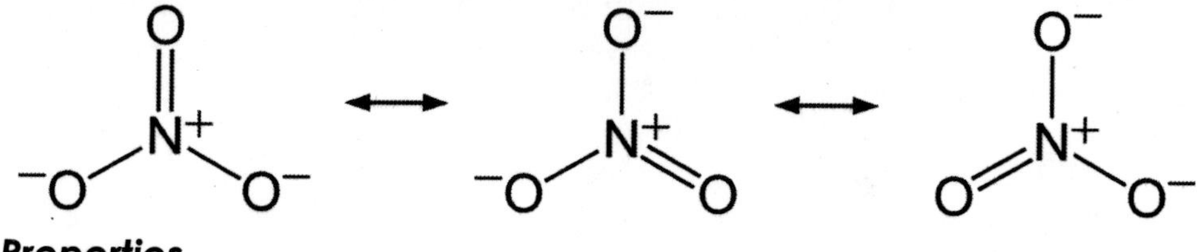

Properties

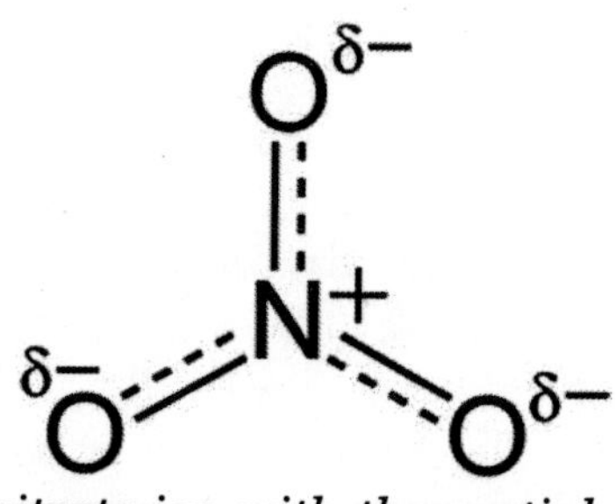

Figure: *The nitrate ion with the partial charges shown*

Almost all inorganic nitrate salts are soluble in water at standard temperature and pressure. A common example of an inorganic nitrate salt is potassium nitrate (saltpeter). A rich source of inorganic nitrate in the human body comes from diets rich in leafy green foods, such as spinach and arugula. It is now believed that dietary nitrate in the form of plant-based foods is converted in the body to nitric oxide which reduces hypertension. Anti-hypertensive diets, such as the DASH diet, typically elevates nitric oxide levels which is first reduced to nitrite in the saliva, as detected in saliva testing, prior to forming nitric oxide.

Occurrence

Nitrate salts are found naturally on earth as large deposits, particularly of Nitratine, a major source of sodium nitrate.

Nitrites are produced by a number of species of nitrifying bacteria, and the nitrate compounds for gunpowder were historically produced, in the absence of mineral nitrate sources, by means of various fermentation processes using urine and dung. Nitrates are found in man-made fertilizers.

Uses

Nitrates are mainly produced for use as fertilizers in agriculture because of their high solubility and biodegradability. The main nitrates are ammonium, sodium, potassium, and calcium salts. Several million kilograms are produced annually for this purpose.

The second major application of nitrates is as oxidizing agents, most notably in explosives where the rapid oxidation of carbon compounds liberates large volumes of gases. Sodium nitrate is used to remove air bubbles from molten glass and some ceramics. Mixtures of the molten salt are used to harden some metals. Explosives and table tennis balls are made from celluloid.

Detection

Free nitrate ions in solution can be detected by a nitrate ion selective electrode. Such electrodes function analogously to the pH selective electrode. This response is partially described by the Nernst equation.

Toxicosis

Nitrate toxicosis can occur through enterohepatic metabolism of nitrate to nitrite being an intermediate. Nitrites oxidize the iron atoms in hemoglobin from ferrous iron (2+) to ferric iron (3+), rendering it unable to carry oxygen. This process can lead to generalized lack of oxygen in organ tissue and a dangerous condition called methemoglobinemia. Although nitrite converts to ammonia, if there is more nitrite than can be converted, the animal slowly suffers from a lack of oxygen.

Phasphorus

Phosphorus is a nonmetallic chemical element with symbol P and atomic number 15. A multivalent pnictogen, phosphorus as a mineral is almost always present in its maximally oxidised state, as inorganic phosphate rocks. Elemental phosphorus exists in two major forms—

white phosphorus and red phosphorus—but due to its high reactivity, phosphorus is never found as a free element on Earth.

The first form of elemental phosphorus to be produced (white phosphorus, in 1669) emits a faint glow upon exposure to oxygen – hence its name given from Greek mythology, Öùóöüñïò meaning "light-bearer" (Latin *Lucifer*), referring to the "Morning Star", the planet Venus. The term "phosphorescence", meaning glow after illumination, originally derives from this property of phosphorus, although this word has since been used for a different physical process that produces a glow. The glow of phosphorus itself originates from oxidation of the white (but not red) phosphorus— a process now termed chemiluminescence.

The vast majority of phosphorus compounds are consumed as fertilisers. Other applications include the role of organophosphorus compounds in detergents, pesticides and nerve agents, and matches.

Phosphorus is essential for life. As phosphate, it is a component of DNA, RNA, ATP, and also the phospholipids that form all cell membranes. Demonstrating the link between phosphorus and life, elemental phosphorus was historically first isolated from human urine, and bone ash was an important early phosphate source. Phosphate minerals are fossils. Low phosphate levels are an important limit to growth in some aquatic systems. The chief commercial use of phosphorus compounds for production of fertilisers is due to the need to replace the phosphorus that plants remove from the soil, and its annual demand is rising nearly twice as fast as the growth of the human population.

Characteristics

Phosphorus exists as several forms (allotropes) that exhibit strikingly different properties. The two most common allotropes are white phosphorus and red phosphorus. Another form, scarlet phosphorus, is obtained by allowing a solution of white phosphorus in carbon disulfide to evaporate in sunlight. Black phosphorus is obtained by heating white phosphorus under high pressures (about 12,000 standard atmospheres or 1.2 gigapascals). In appearance, properties, and structure, it resembles graphite, being black and flaky, a conductor of electricity, and has puckered sheets of linked atoms. Another allotrope is diphosphorus; it contains a phosphorus dimer as a structural unit and is highly reactive.

White Phosphorus and Related Molecular Forms

The most important form of elemental phosphorus from the perspective of applications and the chemical literature is white phosphorus, often abbreviated as WP. It consists of tetrahedral P4 molecules, in which each atom is bound to the other three atoms by a single bond. This P4 tetrahedron is also present in liquid and gaseous phosphorus up to the temperature of 800 °C when it starts decomposing to P2 molecules. Solid white exists in two forms. At low-temperatures, the β form is stable. At high-temperatures the α form is predominant. These forms differ in terms of the relative orientations of the constituent P_4 tetrahedra.

***Figure:** White phosphorus*

White phosphorus is the least stable, the most reactive, the most volatile, the least dense, and the most toxic of the allotropes. White phosphorus gradually changes to red phosphorus. This transformation is accelerated by light and heat, and samples of white phosphorus almost always contain some red phosphorus and accordingly appear yellow. For this reason, white phosphorus that is aged or otherwise impure (e.g. weapons-grade not lab-grade WP) is also called yellow phosphorus. White phosphorus glows in the dark (when exposed to oxygen) with a very faint tinge of green and blue, is highly flammable and pyrophoric (self-igniting) upon contact with air and is toxic (causing severe liver damage on ingestion). Owing to its pyrophoricity, white phosphorus is used as an additive in napalm. The odour of combustion of this form has a characteristic garlic smell, and samples are commonly coated with white "(di)phosphorus pentoxide", which consists of P4O

10 tetrahedra with oxygen inserted between the phosphorus atoms and at their vertices. White phosphorus is insoluble in water but soluble in carbon disulfide.

Thermolysis (cracking) of P_4 at 1100 kelvin) gives diphosphorus, P_2. This species is not stable as a solid or liquid. The dimeric unit contains a triple bond and is analogous to N_2. It can also be generated as a transient intermediate in solution by thermolysis of organophosphorus precursor reagents. At still higher temperatures, P_2 dissociates into atomic P.

Although the term phosphorescence is derived from phosphorus, the reaction that gives phosphorus its glow is properly called chemiluminescence (glowing due to a cold chemical reaction), not phosphorescence (re-emitting light that previously fell onto a substance and excited it).

Red Phosphorus

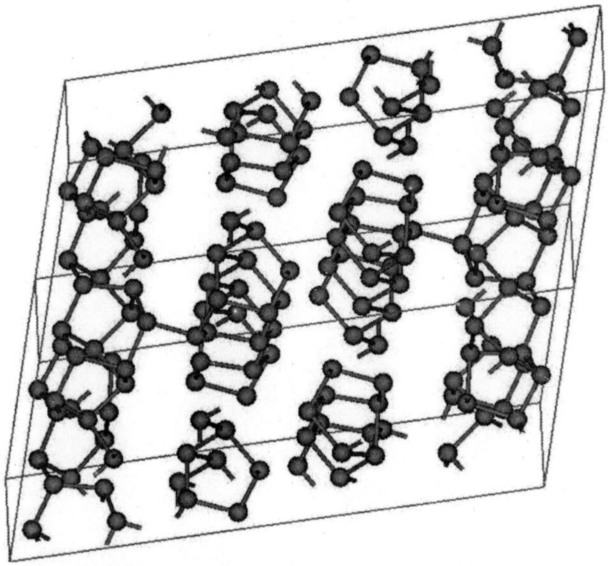

Figure: *Crystal structure of red phosphorus*

Red phosphorus is polymeric in structure. It can be viewed as a derivative of P_4 wherein one P-P bond is broken, and one additional bond is formed with the neighbouring tetrahedron resulting in a chain-like structure. Red phosphorus may be formed by heating white phosphorus to 250 °C (482 °F) or by exposing white phosphorus to sunlight. Phosphorus after this treatment is amorphous. Upon further heating, this material crystallises. In this sense, red phosphorus is not an allotrope, but rather an intermediate phase between the white and violet phosphorus, and most of its properties have a range of values. For example, freshly prepared, bright red phosphorus is highly reactive and ignites at about 300 °C, though it is still more stable than white phosphorus, which ignites at about 30 °C. After prolonged heating or storage, the colour darkens; the resulting product is more stable and does not spontaneously ignite in air.

Violet Phosphorus

Violet phosphorus is a form of phosphorus that can be produced by day-long annealing of red phosphorus above 550 °C. In 1865, Hittorf discovered that when phosphorus was recrystallized from molten lead, a red/purple form is obtained. Therefore this form is sometimes known as "Hittorf's phosphorus" (or violet or α-metallic phosphorus).

Black Phosphorus

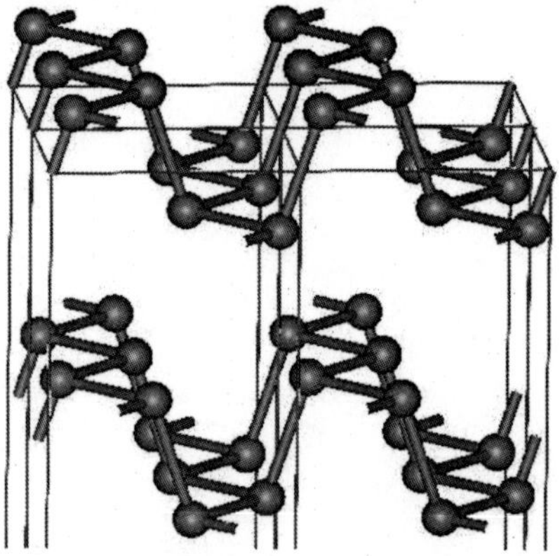

Figure: *Crystal structure of black phosphorus*

Black phosphorus is the least reactive allotrope and the thermodynamically stable form below 550 °C. It is also known as β-metallic phosphorus and has a structure somewhat resembling that of graphite. High pressures are usually required to produce black phosphorus, but it can also be produced at ambient conditions using metal salts as catalysts.

Properties of some allotropes of phosphorus				
Form	***white(a)***	***white(β)***	***violet***	***black***
Symmetry	Body-centred cubic	Triclinic	Monoclinic	Orthorhombic
Pearson symbol		aP24	mP84	oS8
Space group	I43m	P1 No.2	P2/c No.13	Cmca No.64
Density (g/cm³)	1.828	1.88	2.36	2.69
Bandgap (eV)	2.1		1.5	0.34
Refractive index	1.8244		2.6	2.4

Isotopes

Twenty-three isotopes of phosphorus are known, including all possibilities from 24

P up to 46

P. Only 31

P is stable and is therefore present at 100% abundance. The half-integer nuclear spin and high abundance of ^{31}P make phosphorus-31

NMR spectroscopy a very useful analytical tool in studies of phosphorus-containing samples.

Two radioactive isotopes of phosphorus have half lives suitable for biological scientific experiments. These are:

- 32 P, a beta-emitter (1.71 MeV) with a half-life of 14.3 days, which is used routinely in life-science laboratories, primarily to produce radiolabeled DNA and RNA probes, *e.g.* for use in Northern blots or Southern blots. Because the high energy beta particles produced penetrate skin and corneas, and because any 32

P ingested, inhaled, or absorbed is readily incorporated into bone and nucleic acids, Occupational Safety and Health Administration in the United States, and similar institutions in other developed countries require that a lab coat, disposable gloves and safety glasses or goggles be worn when working with 32

P, and that working directly over an open container be avoided in order to protect the eyes. Monitoring personal, clothing, and surface contamination is also required. In addition, due to the high energy of the beta particles, shielding this radiation with the normally used dense materials (*e.g.* lead), gives rise to secondary emission of X-rays via Bremsstrahlung (braking radiation). Therefore shielding must be accomplished with low density materials, *e.g.* Plexiglas (Lucite), other plastics, water, or (when transparency is not required), even wood.

- 33 P, a beta-emitter (0.25 MeV) with a half-life of 25.4 days. It is used in life-science laboratories in applications in which lower energy beta emissions are advantageous such as DNA sequencing.

Occurrence

Phosphorus is not found free in nature, but it is widely distributed in many minerals, mainly phosphates. Historically-important but limited commercial sources were organic, such as bone ash and (in the latter 19th century) guano. Inorganic phosphate rock, which is partially made of apatite (an impure tri-calcium phosphate mineral), is today the chief commercial source of this element. According to the US Geological Survey (USGS), about 50 percent of the global phosphorus reserves are in the Arab nations. Large deposits of apatite are located in China, Russia, Morocco, Florida, Idaho, Tennessee, Utah, and elsewhere. Albright and Wilson in the UK and their Niagara Falls plant, for instance, were using phosphate rock in the 1890s and 1900s from Tennessee, Florida, and the Îles du Connétable (guano island sources of phosphate); by 1950 they were using phosphate rock mainly from Tennessee and North Africa. In the early 1990s Albright and Wilson's purified wet phosphoric acid business was being adversely affected by phosphate rock sales by China and the entry of their long-standing Moroccan phosphate suppliers into the purified wet phosphoric acid business.

In 2012, the USGS estimated 71 billion tons of world reserves, where reserve figures refer to the amount assumed recoverable at current market prices; 0.19 billion tons were mined in 2011. Critical to contemporary agriculture, its annual demand is rising nearly twice as fast as the growth of the human population.

Recent reports suggest that production of phosphorus may have peaked, leading to the possibility of global shortages by 2040. In 2007, at the rate of consumption, the supply of phosphorus was estimated to run out in 345 years. However, some scientists now believe that a "peak phosphorus" will occur in 30 years and that "At current rates, reserves will be depleted in the next 50 to 100 years." Cofounder of Boston-based investment firm and environmental foundation Jeremy Grantham wrote in *Nature* in November 2012, that consumption of the element "must be drastically reduced in the next 20-40 years or we will begin to starve." According to N.N. Greenwood and A. Earnshaw, authors of the textbook, *Chemistry of the Elements,* however, phosphorus comprises about 0.1% by mass of the average rock, and consequently the Earth's supply is vast, although dilute. This book, however, was published in 1997.

Research into Urine Diversion

Urine contains most (94% according to Wolgast) of the NPK nutrients excreted by the human body. Conversely, concentrations of heavy metals such as lead, mercury, and cadmium, commonly found in solid human waste, are much lower in urine (although not low enough to qualify for use in organic agriculture under current EU rules). The more general limitations to using urine as fertilizer then depend mainly on the potential for buildup of excess nitrogen (due to the high ratio of that macronutrient), and inorganic salts such as sodium chloride, which are also part of the wastes excreted by the renal system. The degree to which these factors impact the effectiveness depends on the term of use, salinity tolerance of the plant, soil composition, addition of other fertilizing compounds, and quantity of rainfall or other irrigation.

Urine typically contains 70% of the nitrogen and more than half the phosphorus and potassium found in urban waste water flows, while making up less than 1% of the overall volume. Thus far, source separation, or urine diversion and on-site treatment has been implemented in South Africa, China, and Sweden among other countries with the Bill and Melinda Gates Foundation provided some of the funding implementations. China reportedly had 685,000 operating source separation toilets spread out among 17 provinces in 2003.

"Urine management" is a relatively new way to view closing the cycle of agricultural nutrient flows and reducing sewage treatment costs and ecological consequences such as eutrophication resulting from the influx of nutrient rich effluent into aquatic or marine ecosystems. Proponents of urine as a natural source of agricultural fertilizer claim the risks to be negligible or acceptable. Their views seem to be backed by research showing there are more environmental problems when it is treated and disposed of compared with when it is used as a resource.

It is unclear whether source separation, urine diversion, and on-site urine treatment can be made cost effective; nor whether required behavioural changes would be regarded as socially acceptable, as the largely successful trials performed in Sweden may not readily generalize to other industrialized societies. In developing countries the use of whole raw sewage (night soil) has been common throughout history, yet the application of pure urine to crops is rare. Increasingly there are calls for urine's use as a fertilizer, such as a 2010 *Scientific American* article "Human urine is an effective fertilizer".

Production

The majority of phosphorus-containing compounds are produced for use as fertilisers. For this purpose, phosphate-containing minerals are converted to phosphoric acid. Two distinct routes are employed, the main one being treatment of phosphate minerals with sulfuric acid. The other process utilises white phosphorus, which may be produced by reaction and distillation from very low grade phosphate sources. The white phosphorus is then oxidised to phosphoric acid and subsequently neutralised with base to give phosphate salts. Phosphoric acid obtained via white phosphorus is relatively pure and is the main source of phosphates used in detergents and other non-fertiliser applications.

Elemental Phosphorus

Presently, about 1,000,000 short tons (910,000 t) of elemental phosphorus is produced annually. Calcium phosphate (phosphate rock), mostly mined in Florida and North Africa, can be heated to 1,200–1,500 °C with sand, which is mostly SiO_2, and coke (impure carbon) to produce vaporized P_4. The product is subsequently condensed into a white powder under water to prevent oxidation by air. Even under water, white phosphorus is slowly converted to the more stable red phosphorus allotrope. The chemical equation for this process when starting with fluoroapatite, a common phosphate mineral, is:

$$4\ Ca_5(PO_4)_3F + 18\ SiO_2 + 30\ C \rightarrow 3\ P_4 + 30\ CO + 18\ CaSiO_3 + 2\ CaF_2$$

Side products from this production include ferrophosphorus, a crude form of Fe_2P, resulting from iron impurities in the mineral precursors. The silicate slag is a useful construction material. The fluoride is sometimes recovered for use in water fluoridation. More problematic is a "mud" containing significant amounts of white phosphorus. Production of white phosphorus is conducted in large facilities in part because it is energy intensive. The white phosphorus is transported in molten form. Some major accidents have occurred during transportation, train derailments at Brownston, Nebraska and Miamisburg, Ohio led to large fires. The worst incident in recent times was an environmental one in 1968 when the sea became contaminated due to spillages and/or inadequately treated sewage from a white phosphorus plant at Placentia Bay, Newfoundland.

Another process by which elemental phosphorus is extracted includes applying at high temperatures (1500 °C):

$$2\ Ca_3(PO_4)_2 + 6\ SiO_2 + 10\ C \rightarrow 6\ CaSiO_3 + 10\ CO + P_4$$

Thermphos International is Europe's only producer of elemental phosphorus with the annual capacity of 80,000 t provided by a plant at Vlissingen, Netherlands.

Historically, before technology allowed mineral-based extractions, white phosphorus was isolated on an industrial scale from bone ash. In this scheme the tricalcium phosphate in bone ash is converted to monocalcium phosphate with sulfuric acid:

$$Ca_3(PO_4)_2 + 2\ H_2SO_4 \rightarrow Ca(H_2PO_4)_2 + 2\ CaSO_4$$

Monocalcium phosphate is then dehydrated to the corresponding metaphosphate:

$$Ca(H_2PO_4)_2 \rightarrow Ca(PO_3)_2 + 2\ H_2O$$

Calcium metaphosphate yields, when ignited to a white heat with charcoal, two-thirds of its weight of white phosphorus while one-third of the phosphorus remains in the residue as calcium orthophosphate:

$$3\ Ca(PO_3)_2 + 10\ C \rightarrow Ca_3(PO_4)_2 + 10\ CO + P_4$$

Compounds

Oxoacids of Phosphorus

Phosphorous oxoacids are extensive, often commercially important, and sometimes structurally complicated. They all have acidic protons bound to oxygen atoms and some have nonacidic protons that are bonded directly to phosphorus. Although many oxoacids of phosphorus are formed, only nine are important, and three of them, hypophosphorous acid, phosphorous acid, and phosphoric acid, are particularly important ones.

Oxidation state	*Formula*	*Name*	*Acidic protons*	*Compounds*
+1	HH_2PO_2	hypophosphorous acid	1	acid, salts
+3	H_2HPO_3	phosphorous acid	2	acid, salts
+3	HPO_2	metaphosphorous acid	1	salts
+3	H_3PO_3	(ortho)phosphorous acid	3	acid, salts
+4	$H_4P_2O_6$	hypophosphoric acid	4	acid, salts
+5	$(HPO_3)_n$	metaphosphoric acids	n	salts (n=3,4,6)
+5	$H(HPO_3)_nOH$	polyphosphoric acids	n+2	acids, salts (n=1-6)
+5	$H_5P_3O_{10}$	tripolyphosphoric acid	3	salts
+5	$H_4P_2O_7$	pyrophosphoric acid	4	acid, salts
+5	H_3PO_4	(ortho)phosphoric acid	3	acid, salts

Phosphorus(V) compounds

Oxides

The most prevalent compounds of phosphorus are derivatives of phosphate (PO_4^{3-}), a tetrahedral anion. Phosphate is the conjugate base of phosphoric acid, which is produced on a massive scale for use in fertilisers. Being triprotic, phosphoric acid converts stepwise to three conjugate bases:

$$H_3PO_4 + H_2O \qquad H_3O^+ + H_2PO_4^- \qquad K_{a1} = 7.25\times10^{-3}$$

$$H_2PO_4^- + H_2O \qquad H_3O^+ + HPO_4^{2-} \qquad K_{a2} = 6.31\times10^{-8}$$

$$HPO_4^{2-} + H_2O \qquad H_3O^+ + PO_4^{3-} \qquad K_{a3} = 3.98\times10^{-13}$$

Phosphate exhibits the tendency to form chains and rings with P-O-P bonds. Many polyphosphates are known, including ATP. Polyphosphates arise by dehydration of hydrogen phosphates such as HPO_4^{2-} and $H_2PO_4^-$. For example, the industrially important trisodium triphosphate (also known as sodium tripolyphosphate, STPP) is produced industrially on a megatonne scale via this condensation reaction:

$$2\ Na_2[(HO)PO_3] + Na[(HO)_2PO_2] \rightarrow Na_5[O_3P\text{-}O\text{-}P(O)_2\text{-}O\text{-}PO_3] + 2\ H_2O$$

Phosphorus pentoxide (P_4O_{10}) is the acid anhydride of phosphoric acid, but several intermediates are known between the two. This waxy white solid reacts vigorously with water.

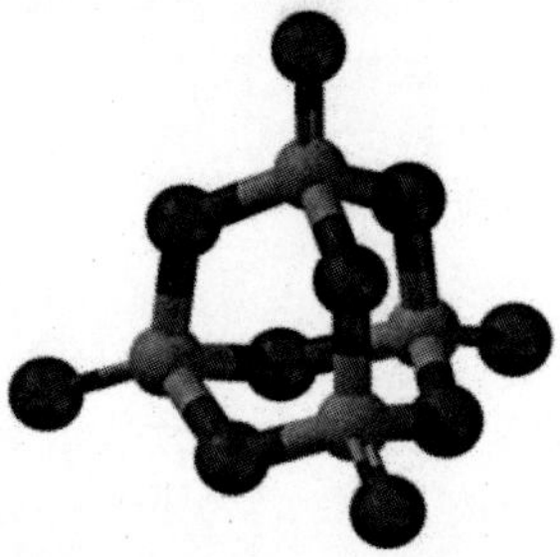

***Figure:** The tetrahedral structure of P_4O_{10} and P_4S_{10}.*

With metal cations, phosphate forms a variety of salts. These solids are polymeric, featuring P-O-M linkages. When the metal cation has a charge of 2+ or 3+, the salts are generally insoluble, hence they exist as common minerals. Many phosphate salts are derived from hydrogen phosphate (HPO_4^{2-}).

PCl_5 and PF_5 are common compounds. Both are volatile and pale or colourless. The other two halides, PBr_5 and PI_5PI_5 are unstable. The pentachloride and pentafluoride adopt trigonal bipyramid molecular geometry and are Lewis acids. With fluoride, PF_5 forms PF_6^-, an anion that is isoelectronic with SF_6. The most important oxyhalide is phosphorus oxychloride ($POCl_3$), which is tetrahedral.

Before extensive computer calculations were feasible, it was proposed that bonding in phosphorus(V) compounds involved *d* orbitals. It is now accepted that the bonding can be better explained by molecular orbital theory and involves only s- and p-orbitals on phosphorus.

Nitrides

Compounds of the formula $(PNCl_2)_n$ exist mainly as rings such as the trimer hexachlorophosphazene. The phosphazenes arise by treatment of phosphorus pentachloride with ammonium chloride: $PCl_5 + NH_4Cl \rightarrow 1/n\ (NPCl_2)_n + 4\ HCl$ The chloride groups can be replaced by alkoxide (RO^-) to give rise to a family of polymers with potentially useful properties.

Sulfides

Phosphorus forms a wide range of sulfides, where phosphorus can be P(V), P(III) or other oxidation states. Most famous is the three-fold symmetric P_4S_3 used in strike-anywhere matches. P_4S_{10} and P_4O_{10} have analogous structures.

Phosphorus(III) compounds

Phosphine (PH_3) and its organic derivatives (PR_3) are structural analogues with ammonia (NH_3) but the bond angles at phosphorus are closer to 90° for phosphine and its organic derivatives. It is an ill-smelling, toxic compound. Phosphine is produced by hydrolysis of calcium phosphide, Ca_3P_2. Unlike ammonia, phosphine is oxidised by air. Phosphine is also far less basic than ammonia.

All four symmetrical trihalides are well known: gaseous PF_3, the yellowish liquids PCl_3 and PBr_3, and the solid PI_3. These materials are moisture sensitive, hydrolysing to give phosphorus acid. The trichloride, a common reagent, is produced by chlorination of white phosphorus:

$$P_4 + 6\ Cl_2 \rightarrow 4\ PCl_3$$

The trifluoride is produced by from the trichloride by halide exchange. PF_3 is toxic because it binds to haemoglobin.

Phosphorus(III) oxide, P_4O_6 (also called tetraphosphorus hexoxide) is the anhydride of $P(OH)_3$, the minor tautomer of phosphorous acid. The structure of P_4O_6 is like that of P_4O_{10} less the terminal oxide groups.

Mixed oxyhalides and oxyhydrides of phosphorus(III) are almost unknown.

Organophosphorus Compounds

Compounds with P-C and P-O-C bonds are often classified as organophosphorus compounds. They are widely used commercially. The PCl_3 serves as a source of P^{3+} in routes to organophosphorus(III) compounds. For example it is the precursor to triphenylphosphine:

$$PCl_3 + 6\ Na + 3\ C_6H_5Cl \rightarrow P(C_6H_5)_3 + 6\ NaCl$$

Treatment of phosphorus trihalides with alcohols and phenols gives phosphites, e.g. triphenylphosphite:

$$PCl_3 + 3\ C_6H_5OH \rightarrow P(OC_6H_5)_3 + 3\ HCl$$

Similar reactions occur for phosphorus oxychloride, affording triphenylphosphate:

$$OPCl_3 + 3\ C_6H_5OH \rightarrow OP(OC_6H_5)_3 + 3\ HCl$$

Phosphorus(I) and Phosphorus(II) Compounds

These compounds generally feature P-P bonds. Examples include catenated derivatives of phosphine and organophosphines. The highly flammable gas diphosphine (P_2H_4) is the first of a series of derivatives of this type. Diphosphine is an analogue of hydrazine. Compounds containing P=P double bonds have also been observed, although they are rare.

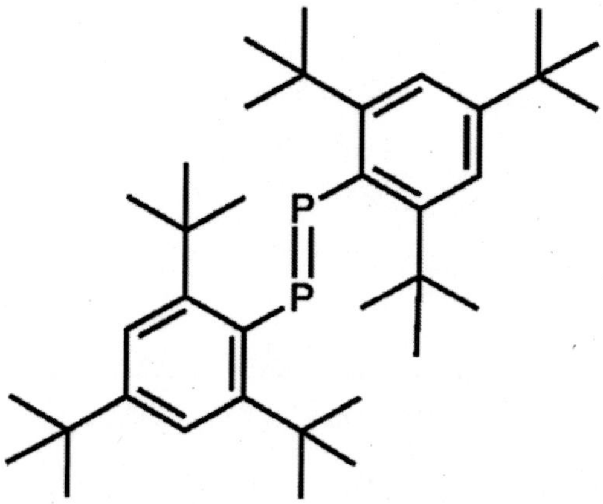

Figure: *A stable diphosphene, a derivative of phosphorus(I).*

Phosphides

The phosphide ion is P^{3-}. Phosphides arise by reaction of metals with red phosphorus. Salts of P^{+} do not exist in solution and these derivatives are refractory, reflecting their high lattice energy. Illustrated by the behaviour calcium phosphide, many metal phosphides hydrolyse in water with release of phosphine:

$$Ca_3P_2 + 6\ H_2O \rightarrow 2\ PH_3 + 3\ Ca(OH)_2$$

Schreibersite is a naturally occurring phosphide found in meteorites. Many polyphosphides are known such as derivatives of OsP_2. These can be structurally complex ranging from Na_3P_7 and derivatives of P_{26}^{4-}. Often these species adopt cage-like structures that resemble fragments of violet phosphorus.

Applications

Fertiliser: The dominant application of phosphorus is in fertilisers, which provides phosphate as required for all life and is often a limiting nutrient for crops. Phosphorus, being an essential plant nutrient, finds its major use as a constituent of fertilizers for agriculture and farm production in the form of concentrated phosphoric acids, which can consist of 70% to 75% P_2O_5. Global demand for fertilisers led to large increase in phosphate (PO_4^{3-}) production in the second half of the 20th century. Due to the essential nature of phosphorus to living organisms, the low solubility of natural phosphorus-containing compounds, and the slow natural cycle of phosphorus, the agricultural industry relies on fertilisers that contain phosphate. A major form of these fertilisers is superphosphate of lime, a mixture of two salts, calcium dihydrogen phosphate $Ca(H_2PO_4)_2$ and calcium sulfate dihydrate $CaSO_4 \cdot 2H_2O$, produced by the reaction of sulfuric acid and water with calcium phosphate.

Widely Used Compounds	***Use***
$Ca(H_2PO_4)_2 \cdot H_2O$	Baking powder and fertilizers
$CaHPO_4 \cdot 2H_2O$	Animal food additive, toothpowder
H_3PO_4	Manufacture of phosphate fertilizers
PCl_3	Manufacture of $POCl_3$ and pesticides
$POCl_3$	Manufacturing plasticizer
P_4S_{10}	Manufacturing of additives and pesticides
$Na_5P_3O_{10}$	Detergents

Organophosphorus Compounds

White phosphorus is widely used to make organophosphorus compounds, through the intermediates phosphorus chlorides and two phosphorus sulfides, phosphorus pentasulfide, and phosphorus sesquisulfide. Organophosphorus compounds have many applications, including in plasticizers, flame retardants, pesticides, extraction agents, and water treatment.

Biological Role

Inorganic phosphorus in the form of the phosphate PO3–4 is required for all known forms of life, playing a major role in biological molecules such as DNA and RNA where it forms part of the structural framework of these molecules. Living cells also use phosphate to transport cellular energy in the form of adenosine triphosphate (ATP). Nearly every cellular process that uses energy obtains it in the form of ATP. ATP is also important for phosphorylation, a key regulatory event in cells. Phospholipids are the main structural components of all cellular membranes. Calcium phosphate salts assist in stiffening bones.

Living cells are defined by a membrane that separates it from its surroundings. Biological membranes are made from a phospholipid matrix and proteins, typically in the form of a bilayer. Phospholipids are derived from glycerol, such that two of the glycerol hydroxyl (OH) protons have been replaced with fatty acids as an ester, and the third hydroxyl proton has been replaced with phosphate bonded to another alcohol.

An average adult human contains about 0.7 kg of phosphorus, about 85–90% of which is present in bones and teeth in the form of apatite, and the remainder in soft tissues and extracellular fluids (~1%). The phosphorus content increases from about 0.5 weight% in infancy to 0.65–1.1 weight% in adults. Average phosphorus concentration in the blood is about 0.4 g/L, about 70% of that is organic and 30% inorganic phosphates. A well-fed adult in the industrialized world consumes and excretes about 1–3 grams of phosphorus per day, with consumption in the form of inorganic phosphate and phosphorus-containing biomolecules such as nucleic acids and phospholipids; and excretion almost exclusively in the form of phosphate ions such as H2PO–4 and HPO2–4. Only about 0.1% of body phosphate circulates in the blood, and this amount reflects the amount of phosphate available to soft tissue cells.

Bone and Teeth Enamel

The main component of bone is hydroxyapatite as well as amorphous forms of calcium phosphate, possibly including carbonate. Hydroxyapatite is the main component of tooth enamel. Water fluoridation enhances the resistance of teeth to decay by the partial conversion of this mineral to the still harder material called fluoroapatite:

$$Ca5(PO4)3OH + F \longrightarrow Ca5(PO4)3F + OH-$$

Phosphorus Deficiency

In medicine, low-phosphate syndromes are caused by malnutrition, by failure to absorb phosphate, and by metabolic syndromes that draw phosphate from the blood (such as re-feeding after malnutrition) or pass too much of it into the urine. All are characterized by hypophosphatemia, which is a condition of low levels of soluble phosphate levels in the blood serum, and therefore inside cells. Symptoms of hypophosphatemia include muscle and neurological dysfunction, and disruption of muscle and blood cells due to lack of ATP. Too much phosphate can lead to diarrhoea and calcification (hardening) of organs and soft tissue, and can interfere with the body's ability to use iron, calcium, magnesium, and zinc.

Phosphorus is an essential macromineral for plants, which is studied extensively in edaphology in order to understand plant uptake from soil systems. In ecological terms, phosphorus is often a limiting factor in many environments; i.e. the availability of phosphorus governs the rate of growth of many organisms. In ecosystems an excess of phosphorus can be problematic, especially in aquatic systems, resulting in eutrophication which sometimes lead to algal blooms.

Food Sources

The main food sources for phosphorus are foods containing protein, although proteins do not contain phosphorus. For example, milk, meat, and soya typically also have phosphorus. As a rule, if one holds a meal plan providing sufficient amount of protein and calcium then the amount of phosphorus is also likely sufficient.

Bibliography

Anon (Inderjit): *Allelopathy, from Molecules to Ecosystems*, Science Publishers, Enfield, New Hampshire, 2002.

Balfour, Lady Eve B. *The Living Soil,* Faber and Faber, London, 1943.

Bennett, Hugh Hammond : *Elements of Soil Conservation*, Biotech Books, Delhi, 2009.

Bolton, Malcolm D.: *A Guide to Soil Mechanics*, Universities Press, Delhi, 2003.

Chandra, Ramesh and Satish Kumar Singh: *Fundamentals and Management of Soil Quality*, Westville Pub, Delhi, 2009.

Dar, Ghulam Hassan: *Soil Microbiology and Biochemistry*, New India Publishing Agency, Delhi, 2010.

Das, Braja M. : *Advanced Soil Mechanics*, Taylor and Francis, Delhi, 2010.

Einhellig, F.A.: *Allelopathy, from Molecules to Ecosystems*, Science Publishers, Enfield, New Hampshire, 2002.

Foth, Henry D. *Fundamentals of Soil Science,* Eighth Edition. New York: John Wylie & Sons, 1990.

Gupta, P.K. : *A Handbook of Soil Fertilizer and Manure*, Agrobios, Delhi, 2011.

Hopkins, Cyril G. *Soil Fertility and Permanent Agriculture*. Boston: Ginn and Company, 1910.

Jose S.: *Chemical Ecology of Plants: Allelopathy in Aquatic and Terrestrial Ecosystems*, Birkhauser Verlag, Basel, Switzerland, 2002.

Macself, R : *Soils and Fertilizers*, Satish Serial Pub, Delhi, 2005.

Majumdar, S.P. and R.A. Singh: *Analysis of Soil Physical Properties*, Agrobios, Delhi, 2008.

Malcolm D. Bolton: *A Guide to Soil Mechanics*, Universities Press, Delhi, 2003.

Mallik, A. U. and Anon: *Chemical Ecology of Plants: Allelopathy in Aquatic and Terrestrial Ecosystems*, Birkhauser Verlag, Basel, Switzerland, 2002.

Mathur, S K B B S Kapoor: *Emerging Trends In Soil Management*, Madhu Publications, Delhi, 2007.

Narwal, S.S. : *Allelopathy In Soil Sickness*, Scientific, Delhi, 2006.

Patel, Sheelwant: *Indian Forests : Soil Water and Bio-Environment Conservation*, Pointer, Delhi, 2005.

Rao, A.V. Narasimha: *Fundamentals of Soil Mechanics*, Laxmi Publications, Delhi, 2010.

Reigosa, M. J., N. Pedrol, A. M. Sanchez-Moreiras, and L. Gonzales: *Allelopathy, from Molecules to Ecosystems*, Science Publishers, Enfield, New Hampshire, 2002.

Russell, E.J. : *Handbook on Soils and Manures*, Discovery, 1999.

Singh, Lalit: *Polymer Microscopy*, London: Hurst and Company Publishers, 2001.

Somani, L L and P C Kanthaliya: *Soils and Fertilisers at a Glance*, Agrotech, Delhi, 2004.

Stone, D. C.: *Instrumental Analysis Laboratory Manual*, Department of Chemistry, University of Toronto, 2005.

Trevor H. : *Transforming Matter: A History of Chemistry from Alchemy to the Buckyball,* Baltimore: Johns Hopkins University Press, 2001.

Vasudev Bhatnagar: *Cell Science and Technology*, Campus Books, Delhi, 2009.

Walters, P.: *The New Quantum Universe*, University Press, Cambridge, 2004.

Willis, R. J.: *Principles and Practices in Plant Ecology: Allelochemical Interactions*, CRC Press, and Boca Raton, FL, 1999.

Index

I

J

K

L

M

N

O

P

R

S

T

U

V

W

Z

□□□